A Welcome
to All Our Visitors

The Royal Observatory, 'the home of time', is one of the most famous historic features of Maritime Greenwich, inscribed by UNESCO as a World Heritage Site in 1997. Situated where east meets west, the Observatory has been at the heart of nautical astronomy since the late seventeenth century, when Charles II commissioned Christopher Wren to design a building in which the first Astronomer Royal, John Flamsteed, could determine the 'longitude of places'.

A £16-million redevelopment and expansion of the Observatory appropriately marks the tenth year since the UNESCO inscription. The new facilities also open exactly seventy years after the formal inauguration of the National Maritime Museum by King George VI. The unchanging task is to show as many people as possible, of all ages and origins, the link between the stars, time and navigation. This work is undertaken in the context of the seafaring history of this fragile planet and investigates distant realms mankind has barely begun to comprehend.

The latest project has totally reconfigured the Royal Observatory site, enabling much greater public access to the buildings as well as to the collections and themes. Among many innovative features for visitors and users are several updated galleries. These include the new Weller Astronomy Galleries and restored Altazimuth Pavilion; modern workshop and research facilities that reveal the richness of the Museum's pre-eminent collection of marine timekeepers; the Lloyd's Register Learning Centre; and the laser-equipped Peter Harrison Planetarium.

The last is housed in a spectacular bronze cone, the shape of which connects astronomical references with earthly latitude and longitude. The Planetarium replaces one first installed at the Observatory forty years ago. It features live astronomy shows and presents a constantly varied programme, both to inspire and inform our ever-growing numbers of visitors from all over the world.

The Museum owes huge thanks to those who have supported the redevelopment, from major public and private funders to thousands of individuals who have made personal contributions. Together, these investments have enabled us to develop the Observatory to its maximum potential and prepare it for another century of service.

We hope you enjoy your visit and that this guide will be both useful and a reminder to come again. Please keep in touch via our website, www.nmm.ac.uk, and do consider becoming a Member of the Museum and receiving regular information about our events and activities.

Roy Clare
Director

The Centre
of Time and Space

The Royal Observatory is the home of the Prime Meridian of the World and of Greenwich Mean Time (GMT).

All time and space is measured relative to the Prime Meridian – also known as Longitude 0° or 000° 00 00 – which is defined by the crosshairs of the great Airy Transit Circle telescope in the Meridian Building of the Observatory.

Greenwich Mean Time is the basis for the International Time-Zone System. How did this small group of buildings become the centre of the world? Who were the people who mapped space and measured time here? What were their tools and the methods they used? What is longitude and why is it measured from an observatory?

This guide not only explores the intriguing history of the Observatory, but also delves into some of the fascinating developments in astronomy and timekeeping in the twenty-first century.

A Guide to the
Royal Observatory, Greenwich
Kristen Lippincott

What is an Observatory?

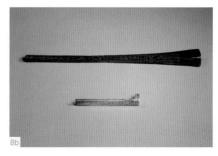

We have evidence for human beings watching and recording the apparent movements of the heavens for over 5000 years. Originally, the earliest astronomers must have been hunters, shepherds and farmers, all watching the skies for signals of the changing seasons – the key to their continued survival. Yet as soon as people began to come together into larger communities, astronomical knowledge became more sophisticated and was often put under the control of rulers and the priestly classes.

An observatory is a place where astronomers watch the apparent movement of the heavens. Most of the early observatories were established in relatively quiet places, set on top of city walls or in towers, where the astronomer-priests would have an uninterrupted 360° view of the heavens. Although none of these very early observatories have survived, echoes of their forms and locations can be seen in the Great Observatory set on the old city walls of Beijing in China, which was built on the site of an earlier observatory; or in the observatory built by Maharajah Jai Singh at Jaipur in Rajasthan, India, which recalls the now-lost observatories of Islamic North Africa and the Middle East.

In antiquity, the earliest systematic observations of the heavens concentrated mostly on the risings and settings of the Sun, the Moon and the stars along the line of the local horizon: hence the need for an unobstructed view. Later, the astronomer-priests of ancient Egypt became more interested in tracking the moment when a certain number of bright stars crossed the highest point in the heavens (the zenith) measured against the north-south line of the local meridian. This marked the beginning of transit instruments and positional astronomy, for which they needed a clear view directly above their heads. In the nineteenth century, however, astronomers turned their eyes towards the whole heavens and designed mountings and machines that could direct their huge telescopes in any celestial direction they wished.

In places where there is relatively little rainfall, such as ancient Egypt and Mesopotamia, observatories would have been open to the skies. In northern climes, however, the astronomer needed a roof to protect him and his instruments from inclement weather. A transit instrument, which observes the movement of a celestial body across a location's meridian, only requires a small slice of the sky to be visible, so its opening to the heavens can be a narrow sliding panel or door. Since the nineteenth century, most large telescopes have been covered with huge, rotatable domes.

The earliest domes, like the original domes at Greenwich, were made of papier mâché, which was not only light and strong but could be easily repaired. These days, most domes are made of fibreglass or high-resistance plastics.

8a. The Ptolemaic universe
The ancient Greeks seem to have been the first to propose a model of the universe in which a central, stationary Earth was enveloped by a series of spheres. They believed that the planets travelled along the surfaces of the inner spheres, and that the outermost sphere was the so-called 'crystalline sphere of the fixed stars'. This understanding of the structure of the universe – the 'Ptolemaic universe', as it came to be known – remained virtually unchallenged for nearly 2500 years. This nineteenth-century cosmosphere preserves part of this idea in having the constellations painted on the surface of a glass celestial globe, with a model of the solar system in the centre. [D7934]

8b. Egyptian astronomy
By combining a sighting device with a scale, the ancient Egyptians created one of the first scientific instruments: the transit instrument like the merkhet shown. By regularly recording the transits of particular stars across the local meridian with an equal-interval timer, such as a water clock, they were able to map the regular movements of the fixed stars with considerable accuracy. The principle of combining a transit instrument with a reliable clock formed the basis for the astronomical measurements made in the Meridian Building at Greenwich.

What is an Observatory?

9a. Stonehenge
The islands of Britain and Ireland are particularly rich in prehistoric archaeological remains. The most famous of these is Stonehenge, the megalithic monument constructed some time between 3000 and 1600 BC. Although the exact significance of this structure is not known, it is unlikely that Stonehenge was designed as an astronomical observatory. Instead, it was probably a place of religious worship that had a specific connection to the apparent movements of the Sun and the Moon. There is ample evidence that many of the stones are arranged along an axis which, in some way, pointed towards either the rising or setting Sun at the time of the midsummer or midwinter solstice.

9b. The oldest observatory
The Chomsung Dae ('near the stars place') Observatory in Kyongju, Korea, is believed to be the earliest surviving observatory in the world. Built in the middle years of the seventh century, it is a simple, beehive-shaped granite structure with a central opening in the roof at the top, from which the astronomer-priests would have observed the transits of the Sun, Moon, planets and stars.

9c. Muslim astronomers
With the end of the classical period, much scientific learning was lost in Europe. In the Islamic world, however, the sciences continued to flourish. From the ninth to the fifteenth centuries, Muslim scholars excelled in mathematical astronomy, which involved systematic observations and mathematical calculations and predictions. This astrolabe is signed by Maḥmūd ibn Shawka from Baghdad and is dated AH694 in the Islamic calendar, which is equal to approximately 1294/95 AD. It can be used both as a calculator and an observing instrument. [E5556-4]

The Crucial Search for Longitude

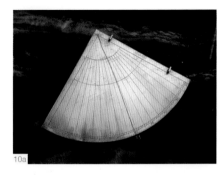

10a.

The story of the Royal Observatory, Greenwich, begins not with the stars but with the sea – and with the most important problem facing all the great maritime nations of the seventeenth century.

One of the great puzzles for scientists throughout the ages was how to measure the exact size of the Earth and how to establish a system for plotting the location of cities and towns on its surface. The ancient Greeks discovered that by measuring the apparent movement of the Sun and the stars over their heads (the science of astronomy), they could calculate the approximate size of the Earth and establish certain coordinates relative to the Equator. These coordinates, measured as angles in degrees north or south of the Equator, are called 'latitude'.

It soon became clear, however, that it was much more difficult to define east/west coordinates, or 'longitude', because there was no fixed point (like the equator) from which to measure. Early civilizations set up a series of arbitrary 'zero points', based on important landmarks or large cities, from which they calculated longitude. But finding longitude on land could only be done by pacing-out distances. There seemed no scientific means for working out how to calculate it.

The difficulty in calculating longitude became particularly serious from the fifteenth century onwards, when increasing numbers of European explorers and traders began to voyage across the seas in search of new lands. It was one thing not to know one's longitude on land; but not being able to calculate it at sea meant that navigators often sailed without knowing how far east or west they were from land. Using a navigational method known as 'dead-reckoning', sailors could make an educated guess about their location by using a compass to check their bearings, and a log to measure the speed of the ship. Nevertheless, cloudy nights or sudden storms at sea often led to shipwreck and tragedy.

10a. Finding latitude
It is relatively easy to determine the latitude of a place – its distance north or south of the Equator. One of the earliest instruments used to do this was the astronomical quadrant, which is simply a quarter of a circle, the curved side of which has been divided into ninety degrees. By sighting an object along one straight edge of the quadrant, the observer can measure the angle of the altitude, or height, of that object by noting where the line of the plumb bob, the weight on the end of the plumb line, crosses the degree scale.

In the northern hemisphere, the angle between the Pole Star, or Polaris, and the horizon indicates local latitude (90° minus the angle in degrees equals the local latitude). In the southern hemisphere, similar measurements can be made using the constellation of the Southern Cross. This brass quadrant dates from about 1600, but it is not known who made it. [B7825]

11a. Sailing by the stars
Even though some early explorers, such as Christopher Columbus, professed to be experts in astronomical navigation, they usually relied much less on their understanding of the movements of the heavens than on their intimate knowledge of the sea. By feeling the shape of the ocean swells, noting the differing degrees of saltiness in the water, checking for flotsam and jetsam and taking samples from the seabed, experienced sailors could read the signs of approaching land. [PU6823]

11b. The sea chart
The earliest surviving sea charts, or *portolani* (literally 'harbour-finders'), date from the middle years of the thirteenth century. The coastlines are drawn with remarkable accuracy, suggesting a long acquaintance with coastal sailing and mapping.

The large expanses of water are covered with a network of criss-crossing lines, known as 'rhumb lines', which radiate from the major ports and specific lines of latitude. These lines all indicate compass directions or bearings. A sailor would set out from port and travel along a particular bearing – say, north-by-north-west – until he crossed a particular latitude or sighted a landmark, when he would change his bearing to maintain his course. Successful navigation depended on having an accurate compass, being able to find the local latitude, and knowing how far the ship had travelled on a given bearing. [D-2150 (detail)]

11c. Telling time by the Bears
Time was kept on board ship in many ways, such as by using a sandglass (hourglass), noting the moment of Noon by measuring the height of the Sun, or using a pocket sundial. At night, sailors could use an instrument called a nocturnal to tell the time. It is a fairly simple device that uses the stars of Ursa Minor and Ursa Major (also known as the 'Big Bear' and 'Little Bear', or the 'Big Dipper' and 'Little Dipper'). The outer edge of the circular part of the instrument has a calendar scale, and the rotating disc within this circle is divided into two periods of twelve hours. To use the nocturnal, the twelve o'clock Midnight mark was lined up against the date. The instrument was held out at arm's length and the Pole Star was sighted through the centre. The pointer was then turned so that it was in line with the 'guard stars' in the Great Bear. The time could then be read off the scale where the pointer cut the hour disc. [D8927]

The Crucial Search for Longitude

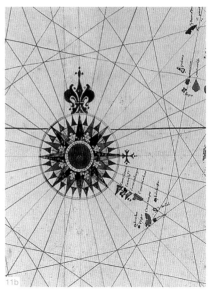

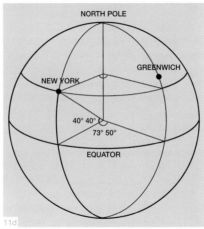

11d. Longitude and time
Both longitude and latitude are measured as segments of a 360-degree circle and are expressed in terms of degrees (°), minutes ( ) and seconds ( ). Since the Earth turns one complete revolution, or 360°, every twenty-four hours, the segments of that circular revolution can be divided into portions of time: 360° equals twenty-four hours; 180° equals twelve hours; 15° equals one hour; and 1° equals four minutes of time.

If you know the difference in local time between two places, you will know the difference in longitude. For example, if a navigator reckons that his local time is five hours different from Greenwich, he knows he is 75° east or west of Greenwich, because 5 x 15° = 75°.

The problem facing navigators well into the eighteenth century was how to know the time in two places without the benefit of a clock that could keep accurate time while travelling on a ship.

11e-11f. The mariner's astrolabe
One of the best navigational tools used by early seafarers was the mariner's astrolabe. It is a simplified version of the astronomical astrolabe. The movable pointer has sights on it and the outer rim of the instrument has a scale. Using these together, the navigator can read the altitude of the Sun or the stars. But taking an accurate reading with such an instrument while standing on the deck of a heaving ship was never easy. [9796 and 8450]

11

Founding the Observatory

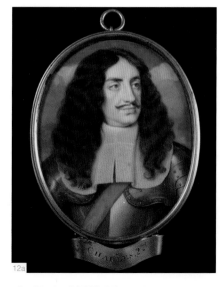

Charles II, King of England ... certainly did not want his ship-owners and sailors to be deprived of any help the Heavens could supply, whereby navigation could be made safer.
John Flamsteed,
***Historia Coelestis Britannica*, 1725**

The European discovery of the Americas and new trade routes to the riches of the East led the foremost maritime nations on a quest for wealth and Empire. The heavy reliance on sea trade and the sheer volume of goods and ships held hostage to the sea meant that finding an answer to 'the Longitude Problem' became an international priority. Not only did every nation want to find the solution, but untold wealth also awaited the nation that could do so first.

In 1674, King Charles II became convinced that there might be an astronomical solution to the Longitude Problem. His mistress, Louise de Kéroualle, Duchess of Portsmouth, had met a Frenchman, a certain Sieur de St Pierre, who claimed that one could plot the motions of the Moon against the field of stars and use the heavens like a big clock in order to determine longitude. Charles II appointed a Royal Commission composed of his most trusted advisors – including Lord Brouncker, President of the Royal Society and Controller of the Navy; Sir Christopher Wren, the King's Surveyor-General, former Gresham Professor of Geometry in London, and Savilian Professor of Astronomy in Oxford; and Robert Hooke, Curator of the Royal Society and Gresham Professor of Geometry – to discover if such a scheme were, indeed, possible. They asked the noted astronomer, John Flamsteed, for his opinion. Flamsteed replied that St Pierre's proposals were ridiculous given the current state of astronomical knowledge: the maps of the stars were inaccurate, there were no reliable tables of the Moon and there were certainly no charts comparing the movements of the Moon against the sphere of the fixed stars.

As a response, Charles II decided that the only way to improve matters was to found an Observatory, and he immediately appointed John Flamsteed as the first Astronomer Royal. Flamsteed's task was

... forthwith to apply himself with the most exact Care and Diligence to the rectifying the Tables of the Motions of the Heavens, and the places of the fixed Stars, so as to find out the so much-desired Longitude of Places for perfecting the art of Navigation.
Royal Warrant, 4 March 1675

12a. *Charles II* (1630–85)
by Samuel Cooper, *c*. 1680
Following fifteen years in exile, Charles II was restored to his father's throne in 1660. Always a strong supporter of the Navy and maritime trade, he established the Royal Observatory by Royal Warrant in 1675. The Royal Society, which was formally constituted under his patronage in 1662, bears further testimony to his scientific interest.

13a. *Louise de Kéroualle* (1649–1734)
by Pierre Mignard, 1682
Born in Brittany, Louise Renée de Penancoët de Kéroualle came to England as part of the entourage of the sister of Charles II, Henriette-Anne, Duchess of Orléans, in June 1670. Within six months, she had become Charles's mistress and, in 1672, she bore him a son, thus securing her status at court as the king's principal mistress. She was created Duchess of Portsmouth in 1673. Louise played an active role in the politics of the day, often using her position to smooth ongoing Anglo-French tensions.

13b. Duke Humphrey's tower
When searching for a site on which to build the Observatory, several options were considered. Sir Jonas Moore, the chief instigator and benefactor of the project, proposed Hyde Park because it was so near to the court. Another suggestion was on the ruins of Chelsea College, near where the Royal Hospital Chelsea is now located. Christopher Wren preferred Greenwich Hill and the site of the recently demolished tower that had been built by Humphrey, Duke of Gloucester (brother of Henry V). During the reign of Elizabeth I, the tower was known as 'Mirefleur', but by the time of the English Civil War (1642–51), when it was garrisoned, it had been much altered and was known as 'Greenwich Castle'. This detail from an etching by Wenclaus Hollar shows it in 1637. [P27662]

Founding the Observatory

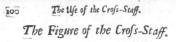

13c. The invention of the telescope

Whereas the magnifying qualities of glass, gemstones and water had been known since antiquity, there is little evidence that they were used to view distant objects until the sixteenth century. The earliest references to the combination of lenses and mirrors to form a primitive reflecting telescope appear in the writings of Leonard Digges of Oxford in 1571, but the first real telescope – a combination of lenses held within a sighting tube – is usually credited to Dutch spectacle-maker Hans Lippershey in 1608. The first astronomer to use the telescope was Galileo Galilei, who made his earliest telescope in May 1609, within twenty-four hours of hearing of Lippershey's invention. By January 1610, he had observed the four moons of Jupiter and the cratered surface of the Moon with his fourth telescope.

When Flamsteed arrived at the new Observatory, he discovered that there were no instruments for him to use. Knowing he would not be able to buy suitable ones from his own meagre salary, he asked for support from his friend and patron, Sir Jonas Moore, who endeavoured to ensure that he was provided with the best technology available. For example, Flamsteed's major observing instruments had telescopic sights, instead of using the older system of pinhole or open sights. The pair also experimented with a long-focus telescope, which pointed vertically upwards and was housed in a 120-foot-deep (40 m) dry well that had been dug in the Observatory grounds. It failed to produce good results on account of the poor quality of its object glass and problems with damp and ground mists.

The telescope shown here is by Christian and Johann Mur in Germany, 1646. This wood and paper telescope is the oldest in the Museum's collection and thought to be the oldest dated lensed instrument in the world. [A5851]

13d. The cross-staff

The cross-staff was originally an astronomical instrument used to measure the angular distance between two objects. There are records of its use as a surveying instrument in 1328 in the writings of French astronomer and mathematician, Rabbi Levi ben Gerson, and it was also simplified and adapted for nautical use by the Portuguese in the fifteenth century. It consists of a rectangular staff, the surface of which has been marked with a degree scale, and two or three perpendicular vanes that can move up and down the staff. Working by the principle of similar triangles, it was useful for measuring the height of the Pole Star above the horizon. [B0038]

Building
Flamsteed House

Whereas, in order to the finding out of the longitude of places for perfecting navigation and astronomy, We have resolved to build a small Observatory within our Park at Greenwich upon the highest ground at or near the Place where the Castle stood, with lodging rooms for our Astronomical Observer and Assistant, Our will and Pleasure is that according to such plot and design as shall be given you by our Trusty and well-beloved Sir Christopher Wren Knight, Our Surveyor General, of the place and site of the said Observatory, you cause the same to be fenced in, built and finished …
Royal Warrant for building the Royal Observatory, 22 June 1675

Charles II instructed Sir Jonas Moore, Surveyor-General of the Ordnance, to begin construction of his Royal Observatory. The foundation stone was set at 3:14 P.M. on 10 August 1675.

The main part of the Observatory was designed by the well-known architect Sir Christopher Wren, who had himself been a professor of astronomy, and the building was set out and supervised by his good friend and colleague, Robert Hooke, who was the City Surveyor for the rebuilding of London after the Great Fire of 1666, as well as Curator of the Royal Society and Gresham Professor of Geometry. The original Observatory had three main floors: a basement area containing a small kitchen, with a small wash-house and workroom nearby; the ground floor with four main rooms, used as a reception hall, study, bedroom and dining room; while on the top floor, the crowning glory of the building was the octagonal 'Star Room'. As Wren recalls in one of his later letters, the building was designed 'for the Observator's habitation and a little for Pompe'.

Among the stipulations, Charles II insisted that the whole building should not cost the Crown more than £500. The work itself was financed through the sale to a Mr Polycarpus Wharton of some old, decayed gunpowder for £1380. Most of the bricks came from Tilbury Fort and some of the wood, iron and lead was taken from a demolished gatehouse in the Tower of London. The exterior was completed by Christmas 1675 and Flamsteed, together with two servants, moved in on 10 July 1676. The total cost to the Crown had been £520 9s 1d.

14a. *John Flamsteed (1646–1719)* **by Thomas Murray, undated**
With the support of Sir Jonas Moore, John Flamsteed was appointed as the first Astronomer Royal – or 'Astronomical Observator' – in March 1675, when he was just twenty-nine years old. He remained in the post until his death in 1719.

15a. The 'birth' of the Observatory
Flamsteed drew this horoscope for the foundation of the Observatory. The annotations are in Latin and read: '1675 Aug [ust] 10 d[ay] 3 h[ours] 14 [= minutes] p.m. [= in the afternoon] lat[itude] 51° 28' 10 .' The line at the bottom of the inside square can be translated as: 'Can you keep from laughing, my friends?' [B0659]

15b. *Flamsteed House, north façade* **by Francis Place, 1676.**
In 1676 Francis Place made some engravings of the Observatory. This print shows the north façade of Flamsteed House. One can see the twin turrets and a telescope mast rising from the roof-line. The largest room of the building is indicated by the long windows of the Octagon Room. The twin summer-houses appear on either side of the building. The eastern (left) one was Flamsteed's solar observatory and the west one (right) was used as a bedroom for his pupils from Christ's Hospital. Behind the house is the mast of Flamsteed's 60-foot (18.5 m) refracting telescope. [7288]

Building Flamsteed House

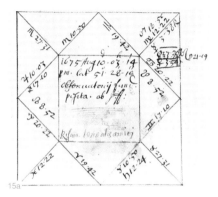

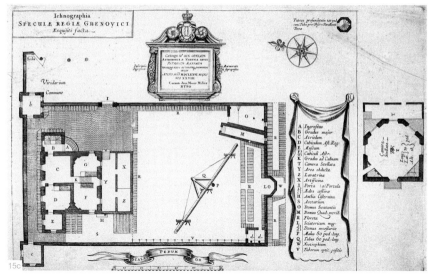

15c. *Flamsteed House*, The original ground-plan by Francis Place, 1676

The original ground-plan of the Observatory is preserved in this etching by Francis Place. As the cartouche in the upper corner tells us, it was *Exquisitè facta*, or 'made with precision' at a scale of 1 inch to 22 feet (2.5 cm to 6.7 m). The room labelled 'D' was the Astronomer Royal's bedroom and 'E' (now subsumed into later enlargements) was his study, or, as the plan calls it, his 'museum' – where the muse of astronomy came to visit him. 'X', 'Y' and 'Z' were covered workrooms and washrooms. Flamsteed grew vegetables in the plots marked 'H' and flowers in the beds marked 'R'. 'W' was a potting shed. The toilets are labelled 'd' and called *domus necessariae*, or 'the house of necessities'. As one can see, the main observing was done in the two buildings at the south end of the garden ('M' housed his meridian quadrant and 'O' his astronomical sextant), the walls of which are aligned north-south, following the line of the local meridian. Finally, in the upper right corner of the print, one sees the circular stair of Flamsteed's well-telescope, which is labelled in Latin: 'Well, 120 feet deep with a tube for observing the parallax of the earth'. [B1652-16A]

15d. The Astronomer's apartments

During the construction of the Observatory, Flamsteed had briefly lived and worked in the Queen's House. The new lodgings were sufficiently complete at the end of May 1676 for him to move some of his equipment up the hill in preparation for viewing a partial solar eclipse on 1 June. On 10 July he moved into the house along with two assistants. One was named Thomas Smith, who was well-qualified for the job; the other, a certain Cuthbert Denton, was 'a silly, surly labourer', whose salary was paid by the Board of Ordnance. The first observation with the Great Sextant was made on 16 September 1676.

For more than 250 years, the Astronomers Royal and their families lived in the ground-floor apartments in Flamsteed House. The original building had only four main living rooms but was extended twice in the eighteenth century and again in 1835–36.

With the outbreak of World War II, most of the occupants of the Observatory were evacuated, including the Astronomer Royal, leaving just a skeleton staff in place to carry out reduced wartime duties. The last full-time occupants of the house were two senior naval officers connected with the Royal Naval College, who left in September 1954. The whole of the Observatory site was then handed over to the Trustees of the National Maritime Museum. Flamsteed House was officially opened to the public by Her Majesty Queen Elizabeth II in July 1960. [D6856]

15e. *Royal Park at Greenwich with a view of the Observatory from the south-west* by an unknown artist, *c*.1680

The Royal Park at Greenwich was created in 1433 but was only occasionally opened to the public before about 1705. This painting by an anonymous artist dates to about 1680 and shows the view of the Park from Croom's Hill. On the left, the walled back garden of the Observatory is visible, with the 80-foot (24 m) mast holding Flamsteed's 60-foot (18 m) refracting telescope. The Queen's House and River Thames can also be seen in the middle of the painting. [BHC1812]

The Octagon Room

The tall windows of the Great Star Room (*Camera stellata*) – or the 'Octagon Room', as it is called today – were designed to accommodate the large telescopes used in the seventeenth century. All these telescopes worked by refracting, or bending, light. Another key instrument used by astronomers was the quadrant. Astronomers fastened a set of sights or a telescope onto a frame in the shape of a quarter-circle, or quadrant. A scale of degrees was engraved on a strip of brass that followed the curved edge of the instrument, thus enabling astronomers to measure the altitude of celestial bodies.

The great panorama of the sky offered by Wren's designs for the Octagon Room meant that the Observatory was perfectly laid out for observing celestial events such as eclipses, comets and planetary movements.

A major feature of the Octagon Room is the pair of 'year-going' clocks built for Flamsteed by Thomas Tompion in 1676. Before Flamsteed could begin his task of charting the stars, he needed to establish whether the Earth rotated at an even rate so that he could have a constant figure as the basis of his measurements. The newly invented pendulum clock provided the first reliable tool with which the rotation of the Earth could be verified.

By the summer of 1676, Flamsteed had reported that the clocks 'kept so good correspondence withe the Heavens' that he was able to prove, to the limit of the technology available to him, that the Earth did indeed rotate at an even rate. From his observations, he had confirmed the data for the 'Equation of Time', which records the changing relationship between Mean Time (that is, 'clock time') and Earth-Sun time. It was not until the invention of the quartz-crystal oscillator in the 1930s that the true irregularity in the speed of the Earth's rotation was discovered.

17a. Fit for a king
The design of the Octagon Room was not only determined by the desire for very tall windows; the Room also needed great height to house the 13-foot (3.96 m) pendulums of Tompion's year-going clocks, which were hung behind the false walnut panelling above the clock movements (the bobs of the pendulums appear in the windows above the two left clock dials). The full-length portraits depict Charles II and his brother, the future James II.
[D7054]

17b. Astronomers at work, 1676
From this engraving by Francis Place, it is clear that the Octagon Room has not changed very much since it was built in 1675. On the left, one astronomer is shown using a quadrant, and on the right, another astronomer uses an 8½-foot (2.5 m) refracting telescope, its viewing angle adjusted by the rungs of a ladder placed against the window. There are records showing that the room was used for observing lunar and solar eclipses, the occultations (or 'obscuring') of stars by the Moon, and for eclipses of the satellites of Jupiter. [8972]

17c. *Sir Christopher Wren* (1632–1723) by T. Smith after Godfrey Kneller, 1711
Wren had been one of the members of the original Royal Commission of 1674, established to examine whether it was possible to determine longitude by astronomical means. Before he turned most of his attention to the rebuilding of London after the Great Fire of 1666, Wren's main interests had been astronomy and mathematics; he had been Gresham Professor of Geometry in London, 1657–61, and Savilian Professor of Astronomy at Oxford, 1661–73. It was also Wren who suggested the site in Greenwich Park as a suitable location for the new Observatory, as it was accessible by road and river but safely distant from the smoky skies of central London.
[PW3176]

17d. Tompion's clock dials
In late 1675, Sir Jonas Moore ordered two clocks for the Observatory from Thomas Tompion, the leading clockmaker working in London. The clocks were installed on 7 July 1676.

The small circular dial at the top indicates seconds, and one revolution of its hand equals a *two*-minute interval. The outermost ring of the main dial indicates minutes, and one revolution of its hand equals *two* hours. The inner ring of the main dial reads hours in the conventional way.
[E7121]

The Octagon Room

17e. The mechanism of Tompion's clock
After Flamsteed's death in 1719, his widow sold the original movements and dials of Tompion's clocks. One appeared at auction in the 1920s and is now in the British Museum. The other belonged to the Earl of Leicester from at least the 1840s and was housed in Holkham Hall, Norfolk. It was brought back to Greenwich in 1994. When the Holkham Hall clock was removed from the Observatory, the movement was adapted for domestic use. This meant replacing the escapement: the part of a clock or watch that corrects and regulates its power of motion. Both Flamsteed and Tompion had agreed that the dead-beat escapement, designed for this clock, was ideal for scientific use because it stopped the wheels – and, therefore, the second-hand – from recoiling backwards at each swing of the pendulum. The loss of the escapement means that our understanding of it remains largely conjectural, resting solely on the brief descriptions and drawings contained in the correspondence between Flamsteed and the distinguished scientist Richard Towneley. Changes to the clock also included the replacement of the long pendulum. The clock is still year-going, but now beats every second, rather than every two seconds, and has a 39-inch-long (1 m) pendulum. [D8930-2]

Early Astronomy at Greenwich

It were much to be wanted our walls might have been meridional but, for saving of Charges, it was thought fit to build upon the old ones which are some 13½° false and wide of the true meridian …
John Flamsteed, 1676

Flamsteed's job was to draw up a map of the heavens sufficiently accurate to be reliable for astronomical navigation. The best way to map the stars is to set up a sighting instrument or telescope along a meridian, or north-south line. As the stars appear to rotate above the astronomer's head, he can measure the position of each star against his meridian, using a timekeeper to help him measure intervals of the Earth's rotation precisely. By comparing thousands of observations from the same meridian, it is possible to build up an accurate map of the night sky.

Unfortunately, since Wren's Observatory had been built on the old foundations of Greenwich Castle, it did not align exactly with the meridian – but headed off slightly to the west. For the purpose of mapping the heavens, Wren's Octagon Room was useless. Instead, Flamsteed set up his positional observatory in a small shed at the bottom of his garden. For the next forty-three years, he worked there, exposed to the night air with the roof shutters drawn back, measuring the transits of the stars over his head.

This small building, housing Flamsteed's 7-foot (2.1 m) astronomical sextant and his 7-foot mural arc, became the real centre of astronomical activity. When Edmond Halley was appointed Astronomer Royal in 1720, he noticed that Flamsteed's brick meridian wall was beginning to subside down the hill into the Royal Park. Halley proposed building a new meridian wall slightly to the north-east of Flamsteed's, upon which he planned to place two new astronomical quadrants. This move eastward established the pattern for the later growth and expansion of the Observatory. Each time a new or better positional telescope was needed, a new room was added to the existing structure, successively further east each time. This series of additions makes up what is now called the Meridian Building.

Walking eastwards from Flamsteed's original meridian, one encounters three later meridian lines: those of Halley, James Bradley and George Biddell Airy – the last of which was recognized in 1884 as the Prime Meridian of the World.

18a. *The Historiae Coelestis Britannica*
In 1704, Prince George of Denmark (consort to Queen Anne) had agreed to pay the entire costs for an edition of Flamsteed's observations. The work went ahead until pressure to publish the material before Flamsteed deemed it ready caused tensions between him and some of Prince George's 'referees'. One of these was Sir Isaac Newton, President of the Royal Society. With the prince's death in 1708, Flamsteed was given some respite, but in 1711 Queen Anne ordered that publication should be resumed.

In 1712, a single volume was published, entitled *Historiae coelestis … observante J. Flamsteedio*. It had been edited by Edmond Halley, one-time close friend to Flamsteed but now Secretary of the Royal Society and his 'sworn enemy'. When Flamsteed saw the new book, he was incandescent with rage. Not only had all but ninety-seven of the sheets been printed without his seeing them; much of the material was not final text, but was only an abridgement of his results. Moreover, sloppy editing had left the text full of errors and he found Halley's sniping preface personally offensive. After repeated pleadings, on 28 March 1716, King George I ordered that 300 of the 400 printed copies of the 1712 edition of the *Historia coelestis* be returned to Flamsteed. He ripped the ninety-seven good pages out of each volume and burned the rest – making a 'sacrifice of them to heavenly Truth'.

Before he died on 31 December 1719, Flamsteed had managed to complete all of volume I (including the ninety-seven good sheets he had saved from the first edition) and part of volume II. Volumes II and III were finally finished and published in 1725 by his wife, Margaret, and two of his former assistants, Abraham Sharp and Joseph Crosthwait. [B3685]

Early Astronomy at Greenwich

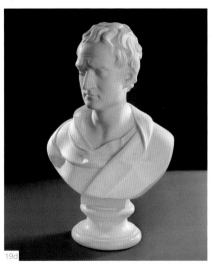

19a. Flamsteed's mural quadrant
An engraving by Francis Place of Flamsteed's 10-foot (3 m) mural quadrant, designed by Robert Hooke. Accurate for its day, the instrument was very difficult to manoeuvre. Flamsteed complained: 'I tore my hands by it'. [A7121(10)]

19b. Flamsteed in later life
As the King's Astronomical Observator, Flamsteed was paid a rather meagre salary of £100 a year. Lack of funds was a constant worry and slowed down his work. To supplement his income, he was forced to teach astronomy to fee-paying pupils (as well as to the two boys from Christ's Hospital he had to train as part of his salary package). The income helped to buy new instruments but he resented the time taken up by his private pupils. 'I must desire to be excused the trouble of them,' he wrote to his patron Sir Jonas Moore, 'since you know well I have work of another nature on my hands.' Finally, in 1685, he was fully ordained and became rector of Burstow, in Surrey, which brought in about an extra £100 a year. This engraving shows Flamsteed at seventy-four years of age in his ecclesiastical robes. It was engraved by George Vertue after the 1712 portrait by Thomas Gibson and was used as the frontispiece to the 1725 edition of Flamsteed's *Atlas*. [A1905]

19c. The Astronomer's study
Although he quarrelled with almost every leading British scientist, Flamsteed still enjoyed great esteem in Europe as a man of learning. He received hundreds of letters from other scientists around the world and was even visited at the Observatory by Tsar Peter the Great of Russia. [D6856]

19d. *Sir Isaac Newton* (1642–1727) by Louis-François Roubiliac, 18th C.
In an age of great scientific minds, Sir Isaac Newton was certainly the leading light. Newton and his scientific colleagues knew that Flamsteed's observations would provide critical data to support their own theories about the structure of the universe. At first, Flamsteed shared his findings quite readily, even seeming to enjoy the camaraderie of a scientific community. But, as pressures to publish his observations mounted, he became increasingly reluctant to release his hard-won findings and became both quarrelsome and isolated from his former colleagues. At one point he even accused Newton of stealing his material – of robbing him 'of the ore' he had worked so hard to dig. When Newton heard this claim, he retorted, 'If he dug the ore, I made the gold ring.' [E8035-1]

19e. *Edmond Halley* (1656–1742) by Sir Godfrey Kneller, *c*. 1721
Despite his battles with Flamsteed, Edmond Halley was appointed the 2nd Astronomer Royal at the age of sixty-four. By this time he was already a famous astronomer, having mapped the stars of the Southern Hemisphere, drawn up charts of the Earth's magnetic variation and accurately predicted the course of the comet that now bears his name. [BHC2734]

The Greenwich Meridian and the Ordnance Survey

James Bradley, the 3rd Astronomer Royal, is known as one of astronomy's most accurate observers. He had made two important discoveries prior to his arrival at Greenwich. Supposedly inspired by the fluttering of a masthead pennant while sailing on the Thames, Bradley was the first astronomer to identify the phenomenon of aberration of light, which is a periodic change in the apparent position of a star caused by the movement of the Earth around the Sun. The discovery of aberration provided real proof of the Copernican theory that the Earth travels around the Sun. It also offered a new way of measuring the speed of light, which Bradley estimated to be about 185,169 miles (298,000 km) per second – only slightly less than the accepted modern value.

Bradley also noticed that that the star *gamma Draconis*, which was often studied by astronomers at Greenwich because it passed directly above the zenith of the Observatory, appeared to change its position in the sky – as much as one second of arc in three days, which was far too much to be due to the aberration of light. After much work, Bradley realized that the apparent change in *gamma Draconis* was actually the result of the Earth wobbling on its own axis due to the uneven gravitational pull of the Moon as it orbits around it. This movement, known as 'nutation', goes through one full cycle every nineteen years. By incorporating corrections for aberration and nutation, Bradley was able to improve the accuracy of the star catalogues significantly.

In 1749, Bradley received money from the Board of Ordnance to build a 'new Observatory' adjacent to Halley's Quadrant Room. Here, he set up his principal telescope, an 8-foot (2.4 m) transit instrument by the celebrated instrument-maker, John Bird. When the French and the British embarked on their great joint project to measure the distance between the Royal Observatories at Paris and Greenwich, the cartographers used the meridian defined by Bradley's new telescope as the official Greenwich Meridian.

Bradley's Meridian was also used as Longitude 0° in the first Ordnance Survey map, one of the County of Kent, published on 1 January 1801. It remained the official Prime Meridian of Britain until 1850, when the 7th Astronomer Royal, George Biddell Airy, decided to build a new transit instrument in the room adjoining Bradley's instrument. To this day, however, all maps produced by the Ordnance Survey still use Bradley's Meridian as their Longitude 0°.

21a. Bradley's Transit Room
James Bradley built his Transit Room in 1749. The instrument in the foreground is the 10-foot (3 m) transit instrument made by Edward Troughton of London in 1816 for John Pond, the 6th Astronomer Royal. This instrument defined the Greenwich Meridian from that date until 1850. [D7061]

21b. *James Bradley* (1693–1762) mezzotint by Johan Faser after John Thomas, 18th C.
James Bradley's first contact with astronomy came through his uncle, who was friends with both Halley and Newton. As with other scientists, Bradley went into the Church, becoming a vicar in 1719 and using his spare time to pursue his interest in astronomy. He became Savilian Professor of Astronomy at Oxford in 1721 and Astronomer Royal in 1742. [PW3410]

21c. The Paris-Greenwich triangulation
In 1743, the French Ambassador in London proposed that it might be useful to produce an accurate map of the area between the Paris and Greenwich meridians. The way to make such a map was by triangulation. A clear baseline is set and, by making measurements through a series of adjoining triangles, very large areas of ground can be surveyed. The English triangulation was carried out by Major-General William Roy, using a chain exactly 100 feet (30.48 m) long as the base measure. This was manufactured by the instrument-maker Jesse Ramsden.

21d. *The Orrery* by Joseph Wright of Derby, c. 1776
One characteristic of the Age of Enlightenment was a profound belief in the beauty of science. People thought of the universe as a great clockwork mechanism, with all its parts working in perfectly synchronized harmony. This painting of a giant orrery, by Joseph Wright of Derby, manages to capture the spirit of the age.

21e. Grand Orrery made by James Simmonds (clock) and Malby & Co. (orrery), 1780.
Mechanical models were a popular means for demonstrating the relative orbits of the planets in the solar system. The orrery is a mechanized planetarium, named after Charles Boyle, 4th Earl of Orrery, who had such a large clockwork model made for him by John Rowley in 1712. [7255]

21f. The constellation of Draco
The constellation of Draco, the 'dragon' or 'serpent', appears to wind around the northern celestial pole. Its third-brightest star *gamma Draconis*, lies just above the serpent's head. [D2614-25]

The Greenwich Meridian and the Ordnance Survey

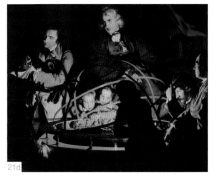

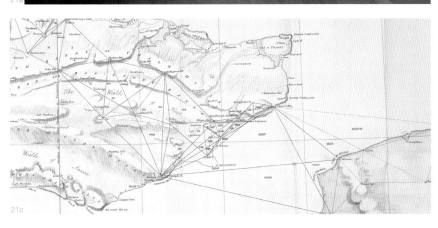

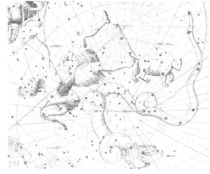

The Golden Age
of Science and Technology

London began to emerge as a centre for the manufacture and trade of precision scientific instruments in the mid-sixteenth century. At that time, numerous land reforms, such as the dissolution of the monasteries under Henry VIII and the beginnings of imperial maritime ambitions under Elizabeth I, meant that there was an increasing demand in England for improved instruments used to measure angles – for surveying, navigation or defence – as well as for improved technology and skills in the engraving and printing processes used to record and disseminate this information.

The skills needed for these trades came from continental Europe, with the most steady stream of immigrants coming to escape the religious turmoil in the Spanish Netherlands, or from one or more of the smaller courts in central and southern Europe. Another element in the growth of the scientific instrument trade was that individual practitioners found it easy to fit in with the established guild structure. Of the twelve great livery companies, the Goldsmiths' had a few instrument-makers on its rolls but it was the Grocers' Company that commanded the lion's share with more than sixty-five master instrument-makers and 200 apprentices listed between 1688 and 1800.

For most of the eighteenth and nineteenth centuries, London was the world centre for making scientific instruments. More often than not, these businesses tended to be family-run enterprises, based on a nucleus of a few skilled craftsmen and their associates. With sophisticated systems of specialized sub-contracting (known as 'part-working') and efficient distribution networks, this was the 'high-tech' industry of its day, providing scientists with the tools they needed to experiment, measure and observe. The Observatory, in particular, benefited from the close working relationship between theorists, practitioners and instrument-makers as it became the proving ground for some of the best technology of the day. In addition, with the main workshops for many of these trades within a short voyage along the Thames, close and ongoing relationships were assured.

Many of the makers started with little formal education, yet all of them – along with the famous scientists whose names are more readily remembered – made a vital contribution to the advancement of scientific knowledge.

22a. Marine sextant by John Bird, c. 1757
John Bird (active 1745–76) was a fine instrument-maker and an inventor. He had worked for both Jonathan Sisson and George Graham, first setting up shop at the Sea Quadrant, near the New Exchange Buildings in the Strand in London. Late in his life he moved to the cleaner air of Little Marylebone Street, with his workshops around the corner in Berwick Street. During this period, he provided several instruments for the Observatory, including a transit instrument and the 8-foot (2.4 m) brass mural quadrant, which remained in use for over sixty-two years. Bird also played a crucial role in the development of the marine sextant, when, together with Captain (later Vice-Admiral) John Campbell, he developed a brass instrument with an arc of 60° that could measure angles of up to 120°.

This brass sextant by Bird shows the very fine quality of his engraving. See also 31e.
[F4928-003]

23a. Astronomical compendium by Humfrey Cole, 1569
Humfrey Cole (c. 1520–91) was one of the finest instrument-makers in Elizabethan England. Earliest records show him working as a die-sinker at the Tower of London Mint but, from this position, he grew to be an extremely sophisticated instrument-maker, able to turn his hand towards the manufacture of any number of different types of instruments. Indeed, he seemed to specialize in trying to multiply the possible uses for any single instrument.

In this compendium, for example, there is equatorial sundial, a nocturnal, a calendar, a circumferentor for surveying, a compass and a geometric square. It can also be used for calculating the times of high and low tides, the phases of the Moon and the height of buildings.
[D8908-A]

The Golden Age of Science and Technology

23a

23b

23c

23d

23b. Portable reflector telescope by Edward Nairne, c. 1770

Instrument-makers showed great pride in the quality of their products. They usually placed their signatures in a prominent location, such as on the clock dial itself or the most visible face of the instrument. On a globe, the signature is usually set within an elaborate decorative cartouche. Often the signature includes the maker's name and address, and sometimes a short paragraph announcing royal or institutional patronage, or advertising other instruments that could be bought from the same maker. Occasionally, however, an instrument is not signed by the maker but by the owner of a shop from which it was sold. Edward Nairne was an important London instrument-maker and a fellow of the Royal Society of London. [C4800/22]

23c. Trade cards

As well as bearing signatures, some scientific instruments have advertisements known as 'trade cards' associated with them. Most often these appear pasted inside the instrument's box or carrying case. Some trade cards are little more than extended signatures; others are often decorative works of art in themselves. Some might show the façade of the maker's shop, provide a series of illustrations of the range of instruments supplied, or show a 'customer' – such as a navigator or a scientist – using one of them. [BD5620-19a]

23d. A family business

As a father-and-son team, the instrument-makers Jonathan Sisson (c.1690–1747) and his son, Jeremiah (1720–83/4), had a workshop at 'The Sphere', at the corner of Beaufort Buildings in the Strand, London. They were incredibly well-connected: Jonathan had worked for clock- and watchmaker George Graham and with the globe-maker Richard Cushee. He also employed John Bird. Jeremiah married a relative of his apprentice James Sidebotham, and employed Jesse Ramsden.

The Sissons are recorded as selling back-staffs, barometers, mural quadrants, octants, rules, sectors, sundials (one of which is shown here), theodolites, transit instruments, globes and micrometers in their shop. As most of their instruments are signed 'J. Sisson, London', separate identification of their work is often very difficult. For the Observatory, however, it is known that Jonathan Sisson assisted George Graham with the construction of Halley's quadrant, and that Jeremiah made two equatorial sectors for Nevil Maskelyne (which were later sent to the observatory at the Cape of Good Hope in South Africa). [D9012]

The Golden Age of Science and Technology

24a

24b

24a. Angle clock by Thomas Tompion, 1691

Thomas Tompion (1639–1713) is documented as having practised from 1671 to 1713. His shop was on Water Lane, near Fleet Street, London, and he was a member of the Worshipful Company of Clockmakers from 1671. His fame and influence were so great that he is often called 'the father of English watchmaking'. He not only provided four completed clocks and part of a fifth one for the Observatory, but also made a number of other scientific instruments, such as barometers and astronomical quadrants. In particular, Tompion enjoyed a close working relationship with both Sir Jonas Moore and Robert Hooke, whose diary is full of references to the projects they worked on together.

This clock is a unique collaboration between Tompion and Flamsteed. Wishing to have the time expressed in a form that he found easy to use, Flamsteed designed a clock that showed time as degrees, minutes and seconds of the arc of the Earth's circle of rotation. In Tompion's clock of 1691, the curved aperture on the right of the dial shows blocks of ten degrees; the central hand indicates individual degrees; the smaller circular dial at the top shows arc minutes and blocks of arc denote seconds.
[D9260]

24b. Astronomical regulator, no.1 by George Graham, 1725

When Edmond Halley arrived at the Observatory in 1720, his first task was to replace all the instruments that had been taken by Flamsteed's widow. He was granted £500 and bought a new mural quadrant and three new clocks from the eminent clockmaker George Graham. Graham's shop was originally in Water Lane, off Fleet Street, London, and he worked on or near Fleet Street for his entire career. He both worked for Tompion and married his niece; and he employed John Bird, Jonathan Sisson, John Shelton and Thomas Mudge. On account of his numerous inventions – including the Graham dead-beat escapement, the mercury-compensated pendulum and the tellurion and planetarium – and his clockmaking skills, he was made a Fellow of the Royal Society in March 1720/1.

This clock, now known as 'Graham 1', cost Halley £12 when it was new. It stood in what is now the Meridian Building, near the stone pillar that Halley constructed to support his new mural quadrant. Later, Graham improved its performance by fitting a temperature-compensating gridiron pendulum, after designs shown to him by John Harrison. [E8946]

25a. Pocket globe by George Adams senior, c. 1770

The Adams were a family of globe- and instrument-makers who were active for nearly 100 years. George Adams senior (1709–72) first opened his shop in Fleet Street, London, on the corner of Racquet Court, in 1734. His business was continued by his son, George junior (1750–95) while his second son, Dudley, first set up his business on the Charing Cross Road in 1788 and then returned to the Fleet Street premises in 1796.

The Adams firm produced a series of optical and mathematical instruments but they are, perhaps, best-known for their pocket globes. First described by Joseph Moxon in 1654, the pocket globe was, essentially, used for the education and amusement of young ladies and gentlemen. It consisted of a small terrestrial globe, usually around 3 inches (7.5 cm) in diameter, fitting into a spherical leather carrying case. Inside the case, there were two hemispherical maps of the constellations.
[D7085-2]

The Golden Age of Science and Technology

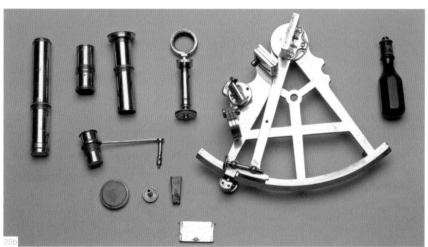

25b. Marine sextant by Jesse Ramsden, c. 1770

Jesse Ramsden came to instrument-making relatively late in life, not applying for apprenticeship until he was twenty-one, bound by a fee of £12. Soon afterwards, though, he acquired a reputation among other makers as a particularly fine engraver of mathematical scales. He began to trade in London under his own name in 1763, first in the Strand, then the Haymarket, and finally at 199 Piccadilly.

The need for accurate scales led Ramsden to develop his first 'dividing engine' in 1767. This could produce precise divisions for all circular instruments such as octants, sextants and theodolites. The scale of a sextant, for example, could be finished in thirty minutes – as opposed to the several hours that a maker needed to divide one by hand. For his development of the dividing engine, as well as 'various inventions and improvements to philosophical instruments', he was given the Royal Society's most prestigious award, the Copley Medal, in 1795. [D6565]

25c. Hand-held telescope by Janet Taylor, 1850

Most instrument-makers were men but there were a few women in the trade. These were most often widows or, sometimes, sisters, carrying on the family business. Mrs Janet Taylor (1804–70) is a notable exception. Daughter of a vicar from County Durham, she arrived in London in 1829, married in 1830 and set up her first shop in Red Lion Street in 1831. In 1835, she established a navigational academy and instrument shop in the Minories called 'The Nautical Academy and Navigation Warehouse'. Though she was supported by her husband, George (and bore eight children, six of whom survived infancy), the business was very much her enterprise. She was a gifted mathematician, patenting a mariner's calculator in 1834 and publishing her own nautical tables. Her general textbook on the art of navigation ran to twelve editions. She was also a regular correspondent with the 7th Astronomer Royal, George Biddell Airy. [F3095]

25d. Selenograph by John Russell, 1797

Numerous astronomers had planned to construct a globe of the Moon, including Johannes Hevelius and Tobias Mayer. The only recorded example, however, was the now-lost early globe made by Christopher Wren in 1661. It is not known what prompted the artist, John Russell R.A., to study the surface of the Moon, but his detailed drawings span a period of over thirty years. His 1797 'Selenographia', as he called it, is not only a beautifully rendered artistic masterpiece but also a unique scientific instrument. It can reproduce and demonstrate all of the various phenomena of the Moon, including libration: the slight changes in the area of its surface visible from Earth. [D7960-A]

A Publick Award:
The Longitude Act, 1714

Regardless of the progress being made at Greenwich to find an astronomical solution to the longitude problem, a series of maritime disasters prompted the British government to seek out alternative means to quicken its discovery. The most notable tragedy occurred on 22 October 1707, when four Royal Navy ships led by Admiral Sir Cloudesley Shovell struck the treacherous ledges off the Isles of Scilly. They all foundered, with the loss of over 1600 lives. Parliament responded to the public outcry by appointing a panel of experts, the Board of Longitude, and in 1714 offered a prize of £20,000 to anyone who could discover a way to determine longitude at sea to within half a degree. They also offered £15,000 for a method to within two-thirds of a degree and £10,000 to within one degree.

As well as attracting serious scientific interest, the Longitude Prize acted as a magnet for many crackpots and their bizarre proposals. Barges moored around the world, all firing flares at Midnight, and perpetual-motion machines sealed in giant vacuum bottles seem sane in comparison to some ideas. One person, for example, proposed using the mysterious 'Powder of Sympathy'. Supposedly, when this powder was sprinkled on a knife which had inflicted a wound on someone, the action would cause the wounded person to re-experience the original pain. It was suggested that if a number of dogs were all wounded with the same knife, they could then be placed on the different ships in His Majesty's fleet. Every day, at Noon, someone at Greenwich could plunge the knife into the Powder of Sympathy and all the dogs would yelp at the same time, regardless of where they were, thus alerting navigators that it was Noon at Greenwich. By knowing the time at Greenwich, they had one essential ingredient towards being able to calculate their longitude at sea. Needless to say, the Board of Longitude was not impressed.

Scientists had long realized that the ideal solution to the longitude problem would be some mechanism that would allow a traveller to know his distance from a zero point (such as Greenwich) in terms of time. In the mid-eighteenth century, however, there simply was no mechanism able to keep good time on a sea voyage, given the heaving motions of the ship and the potential extremes of heat and cold as the vessel travelled from the tropics to the Arctic on voyages of exploration, or in search of trade. Even the great Sir Isaac Newton declared: '… such a watch hath not yet been made'.

26a. *Admiral Sir Cloudesley Shovell* by Michael Dahl, *c*. 1702
Even though Cloudesley Shovell was one of the Navy's most experienced leaders, more than 1600 lives were lost when the ships under his command struck the Gilstone Ledges off the Isles of Scilly. Of the crew of the *Association*, *Eagle* and *Romney*, only one man was saved. Shovell's own body was found floating in the water the next day. [BHC3025]

27a – 27b. The tragedy of the *Centurion*, 1741
In 1740, during growing tensions between Britain and Spain, Commodore George Anson was given command of a squadron of six warships and two storeships to raid and plunder Spanish possessions in the Pacific and gain a foothold on the western coast of the Isthmus of Panama. Narrowly missing the Spaniards on the coast of Patagonia, Anson passed through the Strait of Le Maire and began the westward passage around Cape Horn in March 1741. At that time of year, the weather was dangerous and the westerly currents were strong; Anson's ships were blown 300 miles (483 km) off-course.

Out of sight of land, Anson was not able to find his local longitude. The wandering line on this chart records his desperate attempts: sailing east and then west and then back again, looking for a safe harbour.

When the ships finally reassembled at Más-a-Tierra in the San Fernández Islands in mid-June, the squadron had been decimated. Only Anson's flagship, *Centurion*, completed the full voyage, and although she returned with great booty, the cost in men was also huge. Of 961 who set sail, 626 died from the effects of shipwreck, scurvy, cold and lack of food and water. The log of the *Centurion* records the grisly loss, with 'Dd' ('deceased') indicating the loss of one man after another. [A5477 and A5877]

A Publick Award

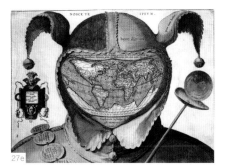

27c. *A fishing boat off a rocky coast in a storm with a wreck* by Jacob Adriaensz Bellevois, 17th C.

Numerous disasters were caused by not knowing longitude at sea. In 1694, Admiral Wheeler's squadron ran aground on Gibraltar when they thought they had already passed through the straits. In 1711, several transport ships were lost near the mouth of the St Lawrence River in North America, when the navigator erred in his calculations by more than 15 leagues (45 miles) in 24 hours. [BHC0837]

27d. The Longitude Act

In May 1714, the House of Commons asked for a scientific evaluation of the various proposals that had been made for 'discovering the longitude' at sea. In the committee's report, dated 11 June, Isaac Newton concluded that four of the proposals had merit and were worth pursuing. To that end, the House presented 'A Bill for Providing a Publick Reward for such Person or persons as shall discover Longitude at Sea'. The first prize was £10,000 for anyone offering a solution to within one degree of a great circle, or sixty geographical miles; £15,000 for two-thirds that distance and £20,000 for determining one-half that distance (half a degree of a great circle, or thirty geographical miles). [D6767]

27e. The folly of the world

By depicting a world map set within the cap of a fool (jester) this satirical print from about 1580 ridicules the imperial ambitions of the great maritime nations. [D2806]

27f. Longitude lunatics

There were so many crazy proposals for solving the longitude problem that 'longitude lunatic' became popular slang for anyone who was deluded. This engraving, part of William Hogarth's series of prints entitled *The Rake's Progress*, depicts the inhabitants of Bedlam, London's notorious insane asylum. One man has a telescope (perhaps for measuring lunar distances); another is shown staring fixedly at a drawing he has scratched on the wall showing a globe marked with lines of longitude and latitude, a ship, the Moon, some geometrical diagrams and a cannon firing a flare.

27g. Dog watch

Today it is difficult to imagine how anyone might have proposed a scheme based on an idea as dubious as the 'Powder of Sympathy'.

We tend to forget that other ages had different assumptions from our own about how the world works. It is easy to laugh but, even in the 21st century, many people believe in the magic of supposedly 'sympathetic' potions and powders. [D6862]

John Harrison:
Finding longitude with a watch

John Harrison was born in 1693, the son of a village carpenter. By the age of twenty, he had taught himself the theory and practical skills of clockmaking and, after the Longitude Prize was announced, Harrison was sure that one of his clocks would win it.

In 1730, following four years of careful thought and study, he had formulated a plan for his first sea going clock. Taking his plans with him, he set off from his home in Lincolnshire for Greenwich to seek advice from Edmond Halley, Astronomer Royal at the time. Halley received Harrison kindly and provided him with an introduction to the greatest clockmaker of the day, George Graham. Graham was entranced by Harrison's plans and even offered him a loan to complete the clock.

Harrison spent the next five years constructing his timekeeper, now known as 'H1'. He then brought it to Graham in London, who arranged to have it publicly displayed to the scientific community. It instantly became quite a celebrity, with several contemporary chroniclers claiming it to be one of the great marvels of the modern age. On its first sea-trials, H1 performed admirably. The Board of Longitude was sufficiently impressed, but Harrison felt that he could better H1's performance and convinced the Board to advance him £250 to make a new version.

Harrison immediately set to work on H2 but soon realized that the machine contained certain deficiencies in its design. He began a third version, H3, and for the next nineteen years obsessively built and rebuilt it. In this he was supported by grants from the Board of Longitude, whose members, it must be admitted, were beginning to lose patience and confidence that Harrison would ever produce the much-vaunted successful marine timekeeper.

The great breakthrough came in 1753, when Harrison commissioned watchmaker John Jefferys to make a small pocket-watch to his own design that he could use to compare the accuracy of his larger timekeepers. As soon as he tested the watch, he realized that he had spent the last twenty-seven years barking up the wrong tree. A small timepiece with a high-frequency oscillator could be fashioned into a much more stable timekeeper than a huge, portable sea-clock would ever be.

29a. The Time and Longitude Gallery
John Harrison's four great timekeepers – H1, H2, H3 and H4 – as well as a number of other important artefacts, are on display in Flamsteed House at the Royal Observatory, Greenwich. The interactive displays provide an encyclopaedia of information about Harrison, his life and his timekeepers. [F5137-008]

29b. Harrison's early wooden clocks
Both John and his brother, James, were interested in clockmaking from an early age. By the mid-1720s, they had already made a series of innovative precision pendulum clocks, which, although relatively ordinary in external appearance, had mechanisms made almost entirely of wood. Only three of Harrison's early wooden clocks and only three of his precision pendulum clocks are known to exist today. This clock from 1713 can now be seen at the Clockmaker's Company Museum at the Guildhall, London. [D6769]

29c. Drawing of escapement and wheels by David Penney, 1993
The upper left section of this illustration by David Penney shows Harrison's invention of a grasshopper escapement, which contained no sliding actions and therefore did not require oiling. The main picture depicts the wooden wheel construction that Harrison used in all his early regulators. [D6733]

29d. H1
H1 was designed to run at a consistent rate, regardless of movement or changes in temperature. Clockwise from the top, the dials show the seconds, hours, a calendar and the minutes. [D6783-3]

29e. H2
Although Harrison himself was never satisfied with H2, it is still a remarkable timekeeper. Completed in 1739, it is larger and heavier than H1. It has improved temperature compensation and a 'remontoir': a mechanism that ensures even drive to the balances. [D6784-2]

John Harrison:
Finding longitude with a watch

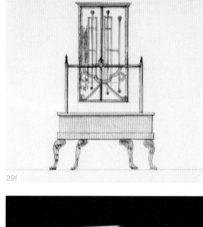

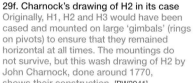

29f. Charnock's drawing of H2 in its case
Originally, H1, H2 and H3 would have been cased and mounted on large 'gimbals' (rings on pivots) to ensure that they remained horizontal at all times. The mountings do not survive, but this wash drawing of H2 by John Charnock, done around 1770, shows their construction. [PW2941]

29g. H3
During the nineteen years that Harrison worked on H3, he introduced many changes in the basic design of the clock. Instead of the dumb-bell balances of H1 and H2, the balances of H3 are wheels. To compensate for temperature changes, Harrison invented the bi-metallic strip, which was made of two flat strips – one of steel and one of brass – that were riveted together. When the bi-metallic strip was heated, the steel lengthened more than the brass and forced the two pieces to curve. When it was cooled, the steel contracted and the strip curved in the other direction. Harrison used the bi-metallic strip to shorten or lengthen the balance spring in order to quicken or slow the clock's motion. The strip is best-known today as an electrical thermostat.
[D6785-5]

Nevil Maskelyne
Finding Longitude by the Stars

While Harrison pursued the idea that one of his timekeepers would solve the longitude problem, others were committed to finding an astronomical solution.

The invention of the reflecting quadrant by John Hadley in 1731 had made the business of measuring celestial angles much more precise. The publication of new lunar tables in 1755, compiled by Professor Tobias Mayer of Göttingen, Germany, meant that the proposal of an astronomical solution to the longitude problem was given new impetus.

Nevil Maskelyne began his astronomical career on 14 July 1760, when he was appointed by the Royal Society to go the island of St Helena to observe the transit of Venus on 6 June 1761. (A transit occurs whenever a small astronomical object passes in front of a larger one; in this instance Venus passed between the Earth and the Sun.) Among his instruments, he had a 20-inch (50 cm) astronomical quadrant, which had been designed and made by John Bird, and a copy of Mayer's solar and lunar tables. Even though the observation of the transit was ruined by cloudy weather, Maskelyne's long voyages to the island and back gave him ample opportunity to conduct a series of experiments using the lunar-distance method of finding longitude at sea.

On his return to England he published *The British Mariner's Guide*, which contained not only an English translation of Mayer's tables but provided simplified instructions for using them, and some sample observations taken during his own travels. Although Maskelyne has recently been cast as the nefarious villain of the John Harrison story, the truth is that, in 1765 – thanks largely to the observations conducted during his trip to St Helena and, later, to Barbados – the lunar-distance method was an increasingly viable option to solving the Longitude Problem. At the important Board of Longitude meeting on 9 February 1765 (coincidentally or not, it was the day after Maskelyne had been appointed Astronomer Royal), both avenues were kept wide open: Harrison was awarded half the amount of the major award (or £10,000) and asked to prove that his timekeeper was, indeed, 'practicable'; while Maskelyne was commissioned to publish an annual almanac containing tables that would facilitate the lunar-distance method for finding longitude.

In 1766, Maskelyne published the first edition of *The Nautical Almanac*, which supplied all the observational data necessary to take a lunar-distance at sea. Using these tables, the able navigator could calculate time at Greenwich from the position of the stars above his head. Once he knew the time at Greenwich, he was halfway towards finding his own longitude at sea.

30a. *Nevil Maskelyne* (1732–1811)
by John Russell, date unknown
From a very early age, Maskelyne knew he wanted to be an astronomer and that the best way to achieve that was to study at Trinity College, Cambridge. One of the requirements for fellowship at Trinity, however, was that he had to take holy orders. While waiting for his application to be cleared, he was ordained and took the curacy of Chipping Barnet in Hertfordshire in 1755, when he was only twenty-three years old. He became a Fellow of the Royal Society less than three years later.

Maskelyne's primary duty was to collate the Observatory's observations for *The Nautical Almanac*. As such, he was entirely responsible for the first forty-nine issues. During his forty-six years as Astronomer Royal, he supervised over 90,000 individual observations, but he also spent a good deal of time arranging for timekeepers to be tested – and was involved in the resulting quarrels with John Harrison, Thomas Mudge, John Arnold and Thomas Earnshaw. In addition, he carried out important experiments to determine the density of the Earth by measuring the deviation of a plumb line produced by the gravitational pull of Schiehallion Mountain in Perthshire, Scotland. He also supervised and planned the scientific sides of the voyages of British explorers James Cook, Constantine Phipps, George Vancouver and Matthew Flinders.
Apart from that, he had a happy home life and was apparently liked and admired by all his colleagues – except, perhaps (as his biographer Derek Howse wryly notes), 'by some chronometer-makers and their families'. In many ways, it is due to Maskelyne's championing of the lunar-distance method and the reliability of his *Nautical Almanac*s that Greenwich is the home of the Prime Meridian of the World today. [F5064]

Finding Longitude by the Stars

31a. The Nautical Almanac
In 1766, the first volume of *The Nautical Almanac* (for 1767) was published, thus completing John Flamsteed's work begun ninety years earlier. It provides tables that list the exact angle between the Moon and certain fixed stars, measured from the Royal Observatory at three-hourly intervals throughout the year. As the path of the Moon varies from year to year, the tables in *The Nautical Almanac* have to be recalculated annually. Despite being slow and complicated to use, its publication cut the arithmetical work that a navigator had to do from four hours to about thirty minutes. [7147]

31b. Tobias Mayer's lunar tables
Mathematicians and astronomers recognized a constant cycle in the relationship between the Sun and the Moon. This so-called 'Saros cycle' ran for eighteen years and 11.3 days, or exactly 223 lunations (the intervals between new moons). Knowing this, one only needed a little over eighteen years of Moon observations and the path of the Moon against the stars would become clear. Halley had started a cycle of observations when he was over sixty years old, so his records were patchy at best, but he was followed in this work by Bradley, who possessed possibly the keenest pair of eyes ever employed at the Observatory. Using his own and Bradley's observations, Tobias Mayer (shown here) of Göttingen, Germany, was finally able to produce a complete set of tables for a whole Saros cycle in 1755. Owing to the constraints of the Seven Years War, which broke out in 1756, Mayer's tables were not fully tested until 1761, during Nevil Maskelyne's long voyage to St Helena to observe the transit of Venus. By using Mayer's observations and a Hadley quadrant, he found he was able to achieve a consistent accuracy of better than 1° – good enough to win the Longitude Prize.

31c. Jupiter's satellites
The earliest longitude prize was actually proposed by Philip II of Spain, who in 1598 offered 6000 ducats as perpetual income, plus 2000 ducats income for life and 1000 ducats for expenses to anyone who could 'find longitude'. Most of the solutions proffered involved magnets and compasses.

In 1616, Galileo Galilei sent his proposal to the Spanish authorities. It was based on the orbiting periods of the moons of Jupiter. As the four largest moons orbit the planet in periods of between 1.75 and seventeen days, they seem to appear and disappear when viewed from Earth. Since Jupiter is so distant from the Earth, these eclipses and occultations seem to occur at the same time, regardless of one's location. In effect, the orbits of the moons of Jupiter provide the perfect celestial timekeeper. Unfortunately, the Spanish refused Galileo's thesis because they felt it would be impossible to make such fiddly observations from the heaving deck of a ship. When the Dutch learned of Galileio's proposals in 1636 and were eager to pursue them, he was under virtual house arrest in Arcetri near Florence and unable to proceed with any negotiations – especially those which were based on changes of appearance in celestial bodies. This was particularly because it was the conclusions he drew from his observation of these changes that had led to his confinement by the Papal authorities in the first place.

31d. Taking a lunar-distance
To find your local longitude, make three nearly simultaneous observations using a watch to time the process. Measure the angular distance between the Moon and a selected star or the Sun (the 'lunar distance'). Measure the altitude of the Moon above the horizon. Measure the altitude of the selected star or the Sun. With this information, you can find the time both at your present location (local time) and at Greenwich (Longitude 0°) by consulting *The Nautical Almanac*. The difference between your local time and the time at Greenwich for the same moment is equal to your longitude – every minute of time difference equals one quarter-degree of longitude. [B4783]

31e. The invention of the sextant
The lunar-distance method requires the navigator to measure the angle between the Moon and the Sun or a star very precisely. Before the 1730s, however, a sufficiently accurate instrument did not exist. In 1731, two men came up with a design almost simultaneously. In England, John Hadley, then vice-president of the Royal Society, proposed a double-reflecting quadrant, which bends the light with two mirrors so that two celestial bodies or a single body and the horizon can be seen simultaneously. At the same time, Thomas Godfrey of Philadelphia, Pennsylvania, also invented a double-reflecting instrument. Unfortunately, the man responsible for lodging Godfrey's claim with the proper authorities did not send his papers to Halley until after Hadley had already published his invention in the *Philosophical Transactions*. And although the Royal Society agreed that there had been a simultaneous discovery of the same principles, all the instruments that were manufactured came to be known as 'Hadley's reflecting quadrant' or, most often, simply as 'a Hadley'. [F4928_001]

H3 + H4:
Harrison's Triumph

But they still say a watch ... can be but a watch ... and the performance of mine (though nearly to truth itself) must be altogether a deception.
John Harrison, 1763

In 1755, Harrison approached the Board of Longitude asking for funds to help him develop H4. Not only were the Board members unsympathetic, but their attention had also been diverted from marine timekeepers towards an astronomical method of finding longitude that seemed much closer to fruition.

H4 succeeded in its first trial, but the Board ruled that it had not been correctly monitored; so, in August 1763, Nevil Maskelyne, who had already developed his own 'lunar-distance method' of calculating longitude and would become the 5th Astronomer Royal in 1765, was sent to Barbados in order to set up an observatory to check the watch's performance when it arrived on its second trial. As it turned out, H4's performance was three times better than that stipulated by the original Longitude Act. But the Board was still not fully convinced that Harrison's watch would truly be 'practicable and useful at sea'. Their real concern was about giving the large Longitude Prize (which could only be claimed once) on the basis of a single timekeeper having performed well on a single trial. The Board knew nothing about how the watch had been constructed, or even if it could be reproduced by anyone other than Harrison himself.

Fairly or not, these uncertainties led the Board to rewrite the terms under which the Prize essentially could be claimed. The new Longitude Act (5, Geo. III, c. 20) became law on 10 May 1765. For Harrison, these changes meant that, in order to claim the £20,000, he would have to divulge the watch's secrets, allow another watchmaker to copy his design, and provide two additional copies of the watch himself. These would then need to be adequately tested before the Board would even consider another request for money. Furthermore, the Board demanded that all four timekeepers be collected from Harrison's home and handed over.

A copy of H4 was duly made, while the watch itself underwent a ten-month trial at the Observatory. According to the records, H4 performed badly. According to Harrison, the trials were a sham: no allowance had been made for the watch's correct rate and the watch had not been adjusted after being dismantled.

Finally, in desperation, Harrison decided to approach King George III directly and request the opportunity to have his next watch, H5, tested by the King himself in his private observatory at Kew. H5 went on trial from May to July of 1772 and, over a ten-week period, its daily rate averaged out at less that a third of a second per day. Harrison approached the Board again, only to be told that the terms of the 1765 Act stipulated that all contenders for the prize had to be tested at the Observatory by the Astronomer Royal and 'no regard will be shewn to the result of any Trial made of them in any other way'.

In April 1773, Harrison petitioned Parliament, asking for justice and laying the blame for everything on the intractable (and unfair) members of the Board of Longitude. Parliament demurred, saying that 'legally', there was nothing it could do. So Harrison drew up a second petition in May 1773, suggesting that the House could pay him a sum of £10,000 directly and, thereby, circumvent the Board altogether. In a very carefully worded document, which received Royal Assent on 1 July 1773, Parliament awarded John Harrison a sum not exceeding £8750 for

having ... applied himself, with unremitting Industry for the Space of Forty-eight Years, to the making of an Instrument for ascertaining the Longitude at Sea ... as a further Reward & encouragement over & above the Sum already received by him for his Invention of a Timekeeper, and his Discovery of the Principles upon which the same was constructed.

The Board did not consider that Harrison's timekeepers had satisfied the requirements of the original 1714 Act but, if one adds up all the money that was ultimately paid to him, the sum actually exceeds the amount of the major award by over £3000. Harrison himself certainly felt vindicated by the prize and died a wealthy man on 24 March 1776 – his eighty-third birthday.

H3 + H4: Harrison's Triumph

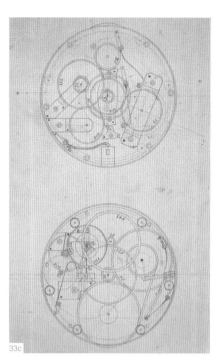

In recalling the event more than twenty-five years later, Nevil Maskleyne admitted that he

> ... always allowed Mr. Harrison's great merit, as a genius of the first rate ... [and] made no opposition to Parliament granting him the remainder of the reward of £20,000; but only to the Board of Longitude doing it; as he [Harrison] had not submitted to trials, and those sufficient to enable the Board to give it to him according to the terms of the Act.

Indeed, the difference between the award that Harrison received and his original request for £10,000 suggests £1250 had perhaps been deducted as a penalty, since Harrison had not complied with the provisions of the Act of 5 Geo. III, c. 20.

Both sides appear to have emerged content from what had been a long and arduous trial. The only ones left to squabble, it seems, are those modern scholars who, even today, continue to disagree as to whether or not Harrison actually won the much-coveted Longitude Prize.

32a. *John Harrison* (1693–1776) by Thomas King, 1767
This portrait of John Harrison shows him holding the Jefferys watch. The cased H3 timekeeper appears behind him to the left. One of his long-case clocks, including Harrison's invention, the gridiron pendulum, appears to the right.
[Science Museum Pictorial/Science & Society Picture Library]

33a. The back plate of H4
Harrison sometimes subcontracted small parts of his timekeepers to other experienced makers. Whereas the inner case of H4 has the mark of 'JH' and the London hallmarks for 1758, suggesting that it could have been made by Harrison himself, the outer case is hallmarked for London 1759 and is stamped, 'HT' – indicating that it probably was made by Henry Thompson, a 'smallworker' and joiner in Silver Street. [D6785-5]

33b. The dial of H4
With its relatively large balance and sophisticated temperature compensations, H4 is the forerunner of all precision watches. It is probably the most important timekeeper ever made. [D0789-A]

33c. Drawing of H4's movement by Harrison, 1760–72
Even though H4 looks like an ordinary pocket-watch, technically it is very different in a number of significant ways. First, it has a very large balance that oscillates five times per second. This means that the balance has more energy stored in it while it is running. The watch also contains a bi-metallic strip, which helps compensate for temperature changes. It has a miniature remontoir, which rewinds eight times per minute in order to keep constant power.

33d. Jefferys' watch
In 1753, Harrison asked John Jefferys, a London watchmaker, to make him a watch following Harrison's own designs. He needed a watch as accurate as possible to help with his astronomical observing and clock-testing. In the end, this watch provided the clue that Harrison needed for improving the design of his timekeepers.

Time for the Navy

... But if Watches made upon Mr John Harrison's or other equivalent Principles should be brought into Use at Sea, the Apparent Time deduced from an Altitude of the Sun must be corrected by the Equation of Time, and the Mean Time found compared with that shewn by the Watch ... as near as the Going of the Watch can be depended upon.
Nevil Maskeyne's 'Explanation' from
The Nautical Almanac **for the Year 1767**

Despite the fact that the numerous trials of John Harrison's timekeepers had proved the possibility of making a reliable longitude watch, his timekeepers were still largely experimental and expensive. Larcum Kendall's first copy of H4, for example, cost the Board of Longitude £500. In order to make this method of finding longitude 'practicable and useful at sea', watchmakers needed to devise a way to manufacture marine timekeepers more quickly and less expensively.

The next chapter in the longitude story was also played out at the Royal Observatory. As a select group of watchmakers created new prototypes, Nevil Maskelyne, the Astronomer Royal, was tasked with arranging and monitoring their trials. Thus began a series of bitter encounters and recriminations as Maskelyne repeatedly found that the watches did not satisfy the terms of the new Longitude Prize and the watchmakers complained that he had not given them a fair trial. While he did not have a vested interest in their failure, as an astronomer, he certainly had no great interest in their success.

From a list of eminent watchmakers, three names stand out. Thomas Mudge had been an expert witness for the Board of Longitude when H4 was being considered. His on-going battles with Maskelyne lasted from 1777 to 1792, with Mudge finally being awarded the sum of £2500. In the end, though, his timekeepers still proved too complex and too expensive for manufacturing on a large scale.

John Arnold probably contributed more to the development of the modern marine timekeeper than any other maker. By simplifying John Harrison's designs, Arnold created the first affordable accurate timekeeper. His watch, made in 1778 and known as 'Arnold, No. 36', was so accurate that it was referred to as a 'chronometer' in 1779 to describe these new, state-of-the-art instruments.

Following Arnold's pioneering work, Thomas Earnshaw's great achievement was to standardize the form of the marine chronometer. He also introduced a series of new manufacturing methods paired with rigid quality controls, which meant his watches could be reproduced quickly, cheaply and in large numbers. By 1800, Earnshaw had created the modern marine chronometer, a design that barely changed in the following 150 years.

With Harrison, Arnold and Earnshaw laying the foundations for an international industry, the chronometer age began in earnest. The British East India Company was quick to demand that all of its ships carry one or two chronometers on board, with the Royal Navy following suit over the next forty years.

And, following in the tradition (if not the spirit) of previous ages, the British Admiralty insisted that each of its chronometers be tested at Greenwich, instituting the annual chronometer trials in 1821.

34a. *Captain James Cook* **(1728-1779) by Nathaniel Dance, 1775**
When Captain James Cook set out on his second Voyage of Discovery (1772–75), he was asked to test K1, Larcum Kendall's copy of H4, and three other marine timekeepers made by John Arnold. Despite the length of the voyage and its extensive travels from Antarctica to the tropics, K1 performed extremely well. The variation in its daily rate never exceeded eight seconds (equal to about two nautical miles at the Equator). As Cook later related to the Secretary of the Admiralty; 'Mr Kendall's Watch has exceeded the expectations of its most zealous Advocate and by being now and then corrected by lunar observations [using *The Nautical Almanac*] has been our faithfull guide through all the vicissitudes of climates.' [BHC2628]

35a. K1
As part of the 1765 stipulations of the Board of Longitude, Larcum Kendall was asked to make a copy of Harrison's H4. Now known as 'K1', it was this watch that was put through such severe trials during Cook's second Voyage of Discovery. As the official astronomer for the voyage, William Wales, commented, '... what an amazing degree of accuracy the ingenious inventor of this watch has brought this branch of mechanics ... let no man boast that he has excelled him, until his machines have undergone a trial as this has done.' [D8550-2]

Time for the Navy

35b. K2
In 1769, the Board of Longitude commissioned Kendall to create a simplified version of H4. He completed the watch, now known as 'K2', two years later, at a cost of £200. In trying to simplify the watch, Kendall left out Harrison's remontoir mechanism. As a result, it never performed very well. [A5510]

35c. *Captain William Bligh* (1754-1817) after John Russell R.A., 19th C.
K2 was issued to William Bligh as he set sail in December 1787 on the *Bounty* to transplant breadfruit from Tahiti to the Caribbean. Following the infamous mutiny in April 1789, K2 was taken by Fletcher Christian and his companions to Pitcairn Island. The last-surviving mutineer, John Adams, sold it to the captain of an American whaling ship in 1808, and by various routes it returned to England in 1840 and then to Greenwich, where it now forms part of the Museum's horological collection. [PU3277]

35d. *The Arnold Family* c.1765
John Arnold was the first watchmaker to tackle the fundamental problem of how to make a commercially viable marine timekeeper. Arnold began experimenting with precision timekeepers in 1767 and supplied three watches for Cook to test on his second Voyage of Discovery. None performed well but they taught him valuable lessons about how they might be improved and simplified. When Arnold's 'No. 36' performed perfectly in its trial at Greenwich in 1779, his supporter Alexander Dalrymple (later Hydrographer of the Navy, responsible for the charting and mapping of seas and oceans) proclaimed that it should be 'named 'chronometer' [since] so valuable a Machine deserves to be known by a Name, instead of a definition'.

35e – 35f. Testing chronometers in the Great Equatorial Building
From the mid-eighteenth century until the 1950s, the testing of marine timekeepers was one of the duties of the Observatory staff. Early watchmakers complained that the conditions in which their watches were tested were less than ideal but, when the Great Equatorial Building was built in 1857, its two lower floors were fitted out to serve as chronometer-testing rooms. Here, the staff checked and rated the accuracy of every single Royal Navy chronometer, often subjecting the timekeepers to extreme heat and cold to test their performance in Arctic or tropical conditions.

From the 1820s, the activities of Admiralty's chronometer-testing service were meticulously recorded in a set of bound ledgers. As a result, it is often possible to recreate a detailed history of each individual chronometer. For example, we have records for Kullberg No. 4070 (the chronometer and log shown here) covering a span of over fifty years. Victor Kullberg began his company in 1851, and throughout the second half of the nineteenth century produced some of the finest chronometers ever made. Kullberg No. 4070 came to the Admiralty trials in Greenwich in 1882. The trials lasted twenty-nine weeks, with 'predictability' (an even rate of loss or gain)

35a

35c

35d

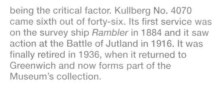
35b

being the critical factor. Kullberg No. 4070 came sixth out of forty-six. Its first service was on the survey ship *Rambler* in 1884 and it saw action at the Battle of Jutland in 1916. It was finally retired in 1936, when it returned to Greenwich and now forms part of the Museum's collection.

The work of horologists at the Royal Observatory has changed over the years. Building on the traditional work of rating timekeepers for accuracy, the team now also looks after the finest collection of precision clocks, watches and chronometers in the world.

Work ranges from regular inspections and lubrication to complete dismantling and the careful cleaning and reinstatement of worn parts. In addition to cataloguing the Museum's collection, the horologists also conduct scholarly research on a wide variety of historical and scientific subjects.
[F6897-001 and F6898]

35e

35f

Time and Society

The term 'Greenwich Mean Time' may be recognized around the world, but in the long history of timekeeping it represents only a small chapter of a much longer story.

Our earliest ancestors used changes in the natural world as the basis of their timekeeping systems. They watched the rising and setting of the Sun, the waxing and waning of the Moon, the passing of the seasons and the movement of the stars across the heavens. By observing and recording these natural rhythms, often using very basic tools, they developed the first notions of timekeeping.

Despite the fact that most cultures base their timekeeping systems on the apparent movements of the Sun, the Moon or the stars (or on combinations of all three), there were never any rules or regulations as to how the calendar should be organized or how the hours should be counted. Instead, most people created structures that best served their particular religion or the climate in which they lived. As a result, even today, people often live their lives according to a 'faith calendar', which runs alongside the local secular calendar. And even though GMT is a world standard, many people still follow timekeeping systems that are more relevant to the way they live their lives. For a farmer, it does not matter what time the clock reads if his animals need to be fed at dawn and dusk. In the far north, where people experience extremely long days in summer and almost perpetual night in winter, it is a continual struggle to align one's natural 'body clock' with the strictly demarcated hours of 'clock-time'.

This variation in how people measure, mark and celebrate the passing of time has created a wonderful variation in the instruments they use. The National Maritime Museum holds a collection of over 6000 time-related instruments, each one different from the next in its design or manufacture.

37a. Astrolabes
The name astrolabe means 'star-taker'. The instrument represents a model of the universe in which the sphere has been flattened. The body of the instrument houses a number of plates, each of which is designed to represent the celestial coordinates for a specific latitude. The pierced covering plate, or 'rete', is a schematic map of the heavens, with each of the pointers representing the position of a bright star. By aligning the rete with the appropriate latitude plate, astronomers could measure the positions of the stars and other celestial objects. Astrolabes could also be used to tell the time. The design of the astrolabe was so good that its shape remained almost unchanged for over 1500 years. This astrolabe was made in Isfahan (Iran) by Muḥammad Khalīl ibn Ḥasan 'Alī and decorated by Muḥammad Bāqir Isfahānī, and is dated (in the Islamic calendar) as AH 1119, which equates to approximately 1707-08 AD. [D7075]

37b. Hourglasses
Some timekeepers do not tell the time of day or night but measure important intervals of time – from a few seconds to several hours. Accurate measurement of this 'interval time' is important in science, sport, business and home life. This set of hourglasses, for example, measures four different intervals of time: fifteen, thirty, forty-five and sixty minutes. The glasses are numbered 1, 2, 3 and 4, to show how many quarter-hours have passed. [E0113]

37c. Perpetual calendars
Most calendars are based on the relative movements of the Earth, Sun and Moon. Unfortunately, these cycles do not keep in exact step with each other. For example, the lunar year is 354 days. The solar year (the orbit of the Earth around the Sun) is 365.242 days. And the sidereal year (the Earth's orbit against the stars) is 365.256 days. These small differences mean that almost all calendars have to be adjusted and updated regularly.

In addition, many religions use both lunar and solar cycles as the basis of their holy days. The Christian calendar is particularly complex as some of its major holy days (such as Christmas) always fall on the same date, while others (such as Easter) have to be recalculated every year. In 325AD, the Council of Nicaea declared that Easter should take place on the first Sunday after the first full moon following the spring equinox (when the path of the Sun intersects the celestial equator). Trying to calculate these four intersecting cycles – of the week, the month, the Sun and the stars – occupied the minds of mathematicians and astronomers for centuries. By the late middle ages, however, the proper cycle had finally been worked out, giving rise to what are known as 'perpetual calendars': calendars that list the day of the week against the calendar date and, especially, the date of Easter for every year 'in perpetuity'. This perpetual calendar appears on the back of a sundial made by Ephraim Senecal in Dieppe, France around 1680. [D8524]

37d. Sundials
The earliest sundials were simple sticks set into the ground. As the Sun passed overhead, the shadows cast by the stick changed in both shape and direction. This simple phenomenon is the basis of the sundial. By dividing the patterns created by the cast shadows into equal segments, the first hour-scales were established.

Most sundials have a pointer called a 'gnomon' which throws a shadow onto a scale to tell the time. The simplest dials have a fixed gnomon and can only be used to tell the time at one particular latitude. Those intended for people wanting to travel have an adjustable gnomon that can be set to a number of latitudes. This dial was made by D. Asselinne in France during the seventeenth century. The two-leaved ivory dial is very finely engraved, and includes a sundial as well as a nocturnal to tell the time at night. [D8842-4]

Time and Society

37a

37b

37c

37d

39a. Setting the time

Early watches were not tremendously reliable, often losing or gaining as much as thirty minutes a day. For this reason, they had to be reset regularly – usually by consulting a sundial. This early multipurpose watch has a built-in sundial and compass. It was probably designed for people who wanted to know the local time regardless of where they were. [4447]

39b – 39c. Wristwatches

Wristwatches were originally a fashion fad for women and were regarded as being too feminine for men. This changed during World War I, when men lost their pocket-watches in the mud during trench warfare. As the nature of trench warfare made timing critical, with troops needing to coordinate going 'over the top' at precisely planned moments, soldiers had to devise a means for keeping hold of their precious timepieces. Some soldiers tied their watches to their wrists, while others set them within a leather strap.

As the demand for wristwatches grew, manufacturers developed a means for making reliable inexpensive watches that would fit anyone's budget. In the 1860s, Roskopf of Switzerland introduced the world's first really affordable watches. In America, manufacturers in the town of Waterbury, Connecticut, began specializing in what they called 'dollar watches' in the 1880s, designed to be cheap to buy and repair. Some even replaced ordinary enamel dials with paper ones in an attempt to bring down costs.

By the end of the twentieth century, nearly everyone wore a wristwatch. Even though some people still prize craftsmanship and pay thousands of pounds for designer watches, most of us now use mass-produced clocks and watches that cost very little. Like this Mickey Mouse novelty watch, made by Ingersoll, USA 1970, and this Bulova tuning-fork wristwatch with transparent Spaceview dial, 1969. [F6398 and F6399]

39d. Astronomical chronometer by George Margetts, 1783

The earliest mechanical timekeepers were the great anaphoric (literally, 'repetitive') clocks of ancient Greece, which seem to have been developed sometime during the third century BC. The purpose of the clock was to provide a two-dimensional account of the daily movement of the Sun and Moon against the background of the fixed stars. This tradition was kept alive in the great medieval and Renaissance astronomical clocks, whose complicated dials showed all sorts of celestial events, such as the phases of the Moon or the positions of the planets in the heavens.

This chronometer by George Margetts is a particularly elegant, eighteenth-century reworking of the astronomical clock, with its main dial representing a planispheric map of the heavens. [D9952]

39e. Novelty clock

As long as there have been clocks and watches, there have been novelty pieces. Some represent advertising or marketing ventures, while others reflect the mood of heartfelt political sentiments. This Chinese clock was made about 1968. With each tick of the clock, the young woman ardently waves her copy of Chairman Mao's 'Little Red Book'.

Today, many young people find they do not need a special watch or clock to tell the time, since they can find the correct time on their mobile telephones. Increasingly, watches are being bought not as timekeepers but as fashion accessories – much as they were in the sixteenth, seventeenth and eighteenth centuries. [D6866]

Time and Society

39a

39d

39b

39c

39e

The Creation of Standard Time

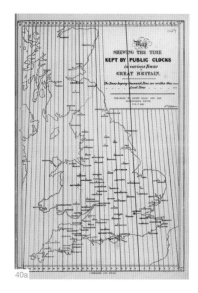

Up until the mid-nineteenth century, every individual town around the world kept local time. There were no national or international conventions on how time should be measured or when the day would begin and end. Some countries, for example, used the system of 'unequal hours', which meant that the length of the hours varied during the year as the balance between daytime and night-time hours changed with the passing seasons. For most, it meant that each town or city had a day made of twenty-four 'equal hours', with the primary fixed point in the day being Noon, the moment when the Sun reached its zenith and crossed the local meridian. This moment was easily measured with a sundial, against which people would set their watches.

Since the Sun passes over any number of meridians in succession as it appears to travel from east to west, the occurrence of 'local Noon' also moves from east to west. When it is noon in London, it is already 12:05 P.M. in Norwich, but only 11:44 A.M. in Plymouth owing to their relative distances east and west of the Greenwich Meridian. Putting it another way, the Sun appears to move 1° of arc across the sky every four minutes, or 15° of arc every hour. Norwich is 1° 15' east of Greenwich and Plymouth is 4° 9' west of it.

Such changes in local time did not really matter until the development of the railway networks. With each town on the railway line following its own local time, the organization of railway timetables became a nightmare. There was one period in the United States when each of the eighty different railway systems kept its own timetable based on the local time of the home depot. A traveller journeying from Maine to California would have to change his watch at least twenty times during the trip in order to be sure of not missing a connection.

These sorts of problems were magnified with the development of the electric telegraph. The first successful marine cable was laid across the English Channel in 1851 and, by 1860, London was connected to the Indian sub-continent by a cable running between Malta and Alexandria. Under the direction of the American Cyrus W. Field, 1879 nautical miles of cable were laid in 1865–66, which connected Ireland to Newfoundland. For the first time in history, virtually every major city could be in direct and immediate contact with the rest of the world. But there was still no internationally agreed system of timekeeping.

40a. British time
As this map from 1852 shows, even across Britain the local time differs significantly from coast to coast. Yarmouth is seven minutes ahead of Greenwich and Penzance is twenty-two minutes behind.

41a – 41b. Double-time
This gold pocket-watch by Benjamin Lewis-Vulliamy was made in about 1847. It has two minute-hands so that its owner might record the local time of his home town as well as that of Greenwich. On the inside of the case, a number of different cities are listed alongside their local difference in minutes from Greenwich Mean Time. [D7165-A, D7165-B]

41c. Railway time
In 1840, the Great Western Railway ordered that 'London time' (that is, Greenwich Mean Time) should be kept at all its stations and in its timetables, but it was thirty years before there was any government action on the matter of legal time for Britain. George Earl's *Perth Station, Coming South*, 1895 shows a clock dominating the busy platform at a railway station.

The Creation of Standard Time

41d – 41e. The Transatlantic Cable

Submarine cables enabled telegraphic signals to be sent across the seas. In 1851, the first successful submarine cable was laid across the English Channel and, by 1860, England and India were connected, with the longest underwater section of 1565 miles lying between Malta and Alexandria.

Establishing communication across the Atlantic Ocean proved much more difficult. After more than two years of trying, the first message was sent on a transatlantic cable – running between Ireland and Newfoundland – on 5 August 1858. The cable had snapped by the following September, however, and was not re-established until 1866. This engraving shows men splicing the transatlantic cable on board the *Great Eastern* in 1865. The medal was produced by Tiffany and Co. in 1858 to commemorate the laying of the cable. In October 1866, time signals sent via the new cable allowed astronomers to redetermine the longitudinal difference between the observatories of Greenwich and of Harvard University in Cambridge, Massachusetts, USA.
[PX8274 and E4502-1]

Greenwich Time for Britain

The Royal Observatory has long been the home of precision timekeeping in Britain. The astronomers' reliance on state-of-the-art timekeeping for the purpose of positional astronomy meant that the best clocks – regulators and marine chronometers – all found their way to Greenwich for testing and approval by the Astronomer Royal.

Whereas the astronomers themselves used several different timekeeping systems depending on which celestial phenomena they were measuring, most of their observations were eventually adjusted to what is known as 'Mean Solar Time'. The Mean Solar Day – or an average solar day based on taking the average of a year's worth of days – was established to provide a more uniform unit of time. Greenwich Mean Time (GMT) is based on the Mean Solar Time as it is measured from the Greenwich Meridian.

As one of the crucial ingredients for navigation at sea was knowing what time it was at Greenwich, the Observatory set a large time ball on the roof of Flamsteed House in 1833 to serve as a visual time signal for all the navigators in the River Thames, who used the ball's daily drop at 1:00 P.M. GMT to calibrate their marine chronometers.

Later, in 1836, the Observatory began to provide the service of 'distributing time' to all the principal chronometer-makers in London. One of the Observatory assistants, John Henry Belville was charged with this responsibility. His wife, Maria, and then later his daughter Ruth Belville, continued it as a private concern. Each Monday morning, they compared the time kept on their large pocket chronometer by Arnold & Son with Greenwich Mean Time, and then set out down Greenwich Hill into the City of London on their rounds. As one contemporary record notes of Ruth Belville,

…she always referred to the watch as Arnold, as if it were the Christian name of a dear friend. Her business with a client would be performed something like this: 'Good morning, Miss Belville, how's Arnold today?' – 'Good morning! Arnold's four seconds fast today', and she would take Arnold from her handbag and give it to you… The [client's] regulator or standard clock would be checked and the watch handed back; and that would be the transaction for the week.

With a growing need for Greenwich to distribute time more widely, the 7th Astronomer Royal, George Biddell Airy, set up a great electrical 'master clock' at the Observatory, which provided impulses via the telegraph system to a number of time signals throughout the nation, twice a day: at 10:00 A.M. and 1:00 P.M. The master clock, designed by London clockmaker Charles Shepherd, was installed in 1852. One of the original 'slave dials' can still be seen outside the Observatory gate.

By the mid-1850s, the necessity for standard time in Britain became the subject of heated debate. Many people, particularly those in the north and west of the country, resented the imposition of time from Greenwich. Since the move towards standard time was spearheaded by the railways, some writers referred to the move as 'railway aggression', thinking that the new proposal was just some clever ploy aimed at the common man by the government and big business. By 1855, however, ninety-eight per cent of all the public clocks in Britain were set to GMT. None the less, it was not until 2 August 1880 that Royal Assent was granted so that 'Wherever any expression of time occurs in any Acts of Parliament, deed, or other legal instrument, the time referred shall, unless it is otherwise specifically stated, be held in the case of Great Britain to be Greenwich mean time…'.

Greenwich Time for Britain

42a – 43a. The Shepherd Gate Clock
This famous landmark is not actually a clock at all but one of the many slave dials that were originally driven by Shepherd's master clock. The dial is unusually laid out with a zero at the top and twenty-three Roman numerals marking the hours in a clockwise fashion.
[B1834 and B4487]

43b. The Greenwich Time Ball
Every day at 12:55 P.M., the Time Ball rises halfway up its mast. It rises to the top of the mast at 12:58 P.M. and then drops exactly at 1:00 P.M. The astronomers chose one o'clock as the time to operate the time signal because, at Noon, they were too busy with their astronomical duties of measuring the transit of the Sun as it passed across the local meridian. [D5600]

43c. Sir George Biddell Airy (1801-92)
Punch, 1883
This caricature from the humorous magazine *Punch* shows the seventh Astronomer Royal, Sir George Biddell Airy, as the Greenwich Time Ball. The Moon looks very amused by the proceedings to standardize the passage of time. [D7065]

43d. Ruth Belville
Miss Ruth Belville was the 'Greenwich Time Lady' from 1892 until the 1930s. She called at forty to fifty establishments every week with her Arnold chronometer, No. 485/786. Here, she is shown having her time card verified in front of the Shepherd Gate Clock.

43e – 43f. Distributing time
In the nineteenth century, a number of time balls were added to the skylines of British towns. The Charing Cross time ball on the Strand was erected in 1852, and the time ball in Deal in 1855. The Calton Hill time ball in Edinburgh was installed in 1854. Similar technology was also used to control the time guns at Edinburgh Castle from 1861 and at Dover Castle from the 1920s onwards. Smaller time signals were also set up in the windows of jewellers to allow the public access to Greenwich Time as they passed by.
[C2985 and C2986]

Greenwich Time
for the World

The need for standard time in Britain, where the maximum difference in longitude equals less than thirty minutes of time, was mild compared to the problems facing Canada and the continental United States, where the time difference between the east and west coasts added up to more than three and a half hours. Professor Charles Ferdinand Dowd was the first to propose that both countries adopt a shared time-zone system, whereby every 15° of longitude would equal one hour's worth of time, and the time would be uniform across each 15° zone.

Dowd's proposal went into effect in the United States in 1883, with the peculiar wrinkle that Greenwich – rather than an American or Canadian city – was chosen to serve as Longitude 0° for the whole system. Initially, Dowd had proposed the US Naval Observatory in Washington DC as the zero-point, but he appears to have run into a bit of trouble when he made the proposal to a Boston-based railway company in 1871. The choice of Greenwich may have been the result of a desire to circumvent local rivalries.

At the same time, discussions were taking place in various world capitals about the possibility of establishing a time-zone system for the whole world. As one might expect, the main sticking point was which nation would be accorded the honour of being the home of the 'Prime Meridian of the World'.

In October 1884, forty-one delegates from twenty-five nations convened in Washington DC for the International Meridian Conference. By the end of the proceedings, seven important principles had been voted through:

- It was desirable to adopt a single world Prime Meridian in place of the innumerable meridians that existed.
- The meridian passing through the principle transit instrument of the Observatory at Greenwich was to be this 'initial meridian'.
- All longitude would be calculated both east and west from this meridian up to 180°.
- All countries would adopt a universal day.
- The universal day would be a Mean Solar Day, beginning at the moment of Mean Midnight at Greenwich and counted on a twenty-four-hour clock.
- Nautical and astronomical days everywhere would begin at Mean Midnight.
- All technical studies designed to regulate and extend the application of the decimal system to the division of time and space would be supported.

Greenwich was named as Longitude 0° by a vote of twenty-two in favour to one against (San Domingo), with two abstentions (France and Brazil).

There were two main reasons for the victory. The first was the fact that the United States had chosen Greenwich as the basis of its own national time-zone system. The second, pointed out by Sandford Fleming, the British delegate representing Canada, was that if one calculated the total tonnage of vessels sailing the seas, seventy-two per cent of the world's commerce depended on sea-charts that used Greenwich as the Prime Meridian. The decision, essentially, was based on the argument that, by naming Greenwich as Longitude 0°, it would inconvenience the least number of people.

Greenwich Time for the World

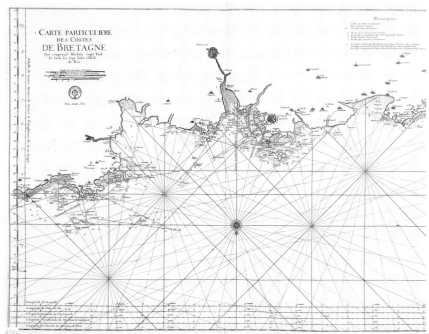

45c. Charles Dowd (1825–1904)
In 1870, Charles F. Dowd published a 107-page pamphlet entitled *A System of National Time for Railroads*, in which he proposed that the continental United States and Canada should be divided by four meridians into five 15°-wide zones in which the time would be uniform. Later, it was agreed that the meridians could be moved slightly east or west to follow local state or county boundaries. For some cities located on or near a border of two time zones, the choice of which time to adopt became intensely political. Detroit, Michigan, for example, which sat squarely on the borderline between the Eastern and Central zones, flip-flopped for over sixteen years between the two – the whole city using local time until 1900, when half the town switched to Central Standard Time; then the whole of the town moved to CST in 1905; and then the whole town switched over to Eastern Standard Time in 1915, though the move was not voted into law until 1916.

45a. The International Meridian Conference, 1884
Most members of the International Meridian Conference were professional diplomats, though some countries sent scientific and technical representatives as well. In the official 166-page report of the discussions, more than seventy-one pages are devoted to the resolution of Greenwich as the Prime Meridian. Curiously, the next most contentious subject was the adoption of a twenty-four-hour day beginning at Mean Midnight.

45b. The problem of multiple longitudes
This detail from one of the official French atlases, *Le Neptune françois*, shows the problem a sailor would encounter when sailing in international waters. At the bottom of the chart, there are five different longitude systems that might be used off the coast of Britain – based on meridians at Tenerife, Ferro Island (El Hierro, in the Canary Islands), The Lizard (Cornwall), London (Greenwich) and Paris. [B5757]

What Time Is It – Today?

In the twentieth century, radio signals became an increasingly important means of time distribution. This provided an extra method by which the accuracy of chronometers could be checked at sea. The earliest wireless time signals were broadcast by the United States Navy in the spring of 1904. Regular services were set up in France and Germany in 1910. Oddly, perhaps, Britain delayed setting up its own system. Instead, a room was fitted out in the Observatory to listen to the wide array of foreign signals in order to collate them and compare their accuracy against Greenwich Mean Time, and to report discrepancies whenever they occurred.

By 1911, it became clear that the time signals from different wireless stations could vary from each other by several seconds. Taking the initiative, the French offered the services of the Paris Observatory to establish an international time association. The *Bureau International de l'Heure* (or 'BIH', as it was often known), began operation nine years later, on 1 January 1920.

With the introduction of quartz-crystal technology, it soon became apparent that these new clocks were 'more accurate than the Earth'. That is to say, if one compared the length of a Mean Solar Day with the equivalent number of seconds as measured by the new clocks, it was evident that the Earth did not rotate evenly on its axis, as Flamsteed had shown, but that there were at least three kinds of irregularities: (1) secular changes, or the gradual slowing down of the planet by about 1.7 milliseconds per century, due to friction and changes in the shape of the Earth itself; (2) irregular fluctuations, caused by the differing rates of rotation of the Earth's molten core and its surface, which causes the overall rotation of the planet to speed up or slow down by up to four milliseconds per decade; and (3) seasonal variations, caused by changes in the patterns of the winds and other weather effects.

By 1950, it was agreed that something needed to be done – but what? Should the concept of the Mean Solar Day as a fundamental unit of time be abandoned? Or should the clocks just be adjusted as required, thereby reasserting that the relationships between the Sun, Moon, Earth and stars are the real arbiters of time?

As this guide goes to press, there are a number of accepted conventions for telling the time. The main ones are:
- GMT, also known in the past as 'Universal Time' (UT), is based on the spin of the Earth on its axis. GMT is still recognized as legal time in Britain and world-wide as the accepted 'short hand' term to indicate international standard time.
- UT 0 is Mean Solar Time at the Prime Meridian obtained from direct astronomical observation. UT 1 is UT 0 corrected for observed polar motion (the Earth wobbling on its axis, causing about 0.035 seconds of variation per year). It is the time scale used for celestial navigation. UT 2 is UT 0 corrected for observed polar motion and extrapolated variations of the Earth's rate of rotation.
- TAI is *Temps Atomique International* (International Atomic Time), in which each second is defined as 'the duration of 9,192,631,770 periods of the radiation corresponding to the transition between two hyperfine levels of the ground state of the caesium-133 atom'.
- UTC, or 'Coordinated Universal Time', was developed in 1958 in an attempt to keep Universal Time (UT) and Atomic Time (TAI) to within approximately one-tenth of a second of each other.

Every few years, one country or another suggests that we should abandon UTC altogether and move to a purely atomic-clock-based timescale. Were that to happen, the time on Earth would no longer be based on our relationship with the Sun, Moon and stars, and a tradition stretching back millennia would end.

What Time Is It – Today?

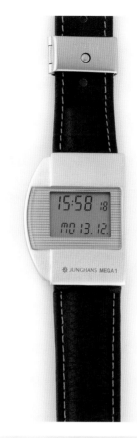

47a. Quartz-crystal clocks
In the 1920s, a group of telecommunications engineers began to experiment with materials that might provide a reliable standard frequency for electromagnetic waves. Quartz crystal was soon identified as a highly stable frequency source for radio communications and, by 1928, there was a viable proposal for incorporating quartz-crystal oscillators into a clock. The first quartz clock to be installed in Greenwich was set up in 1939. It had an accuracy of about nine milliseconds per day. This quartz clock dates to around 1944 and was built at the General Post Office research station in London. [F4517-002]

47b. Atomic time
Although much better than all mechanical clocks, the quartz clock tended to suffer from 'drift' as its crystals aged. The next material used was the ammonium atom, with the first complete operational atomic clock developed at the United States Bureau of Standards in Washington DC by Harold Lyons and his associates in 1948–49. But it was at the National Physical Laboratory (NPL) in England that a caesium-133 atomic clock was first used on a regular basis, with a service beginning in June 1955 and leading to the redefinition of the second in atomic terms in 1967. This particular caesium-133 atomic clock was made in California by Hewlett Packard in 1991 and was used at NPL during the early 1990s.

47c. Radio-controlled watches
Coded radio time signals were first used by the American military in 1904. The first civilian system was begun in 1910, when the French started to transmit signals from the Eiffel Tower in Paris. But it was not until the 1970s that microelectronics enabled clocks to set themselves automatically from signals from different transmitters around the world. The first commercially produced radio-controlled wristwatch was manufactured by the German firm of Junghans. In Germany, such clocks and watches receive the radio signals from Frankfurt, while English versions pick up the signals from the NPL's radio transmitter at Anthorn, Cumbria.

Even though the time services left Greenwich in 1940, the term 'Greenwich Mean Time' continues to be used. In fact, the first atomic time measured at the NPL was actually known as 'GA' or 'Greenwich Atomic' and was used internationally until about 1959. Even today, all the forms of Universal Time (UT) are still based on measurements taken from the Prime Meridian at Greenwich. The occasional adding of leap seconds is, in effect, bringing all these systems back into line with the defining coordinates of the Airy Transit Circle. Today, even though everyone knows and uses the initials GMT, the most widely used time system is UTC. [F3926-2]]

The *Camera Obscura*

The principle of a *camera obscura* was first fully demonstrated by the Greek mathematician Euclid in about 300 BC. If the wall of a darkened room (*camera obscura* in Latin) is pierced with a hole so tiny that only a concentrated beam of light can enter, the light will project an inverted image of the outside world onto the opposite wall. The effect seemed magical to the ancient world, but the practicality of its application soon became apparent. First, it provided a very effective way for those in power to spy on people without being observed. By the eleventh century, however, the Arabic scientist Ibn al-Haytham (usually known in the West as 'Alhazen') realized that the device could be used to observe solar phenomena, such as eclipses, without burning one's eyes. Using Alhazen's ideas, William of Saint-Cloud published a set of instructions on how to build and use a *camera obscura* for astronomical work in 1290. Lenses were added to the device in the sixteenth century and portable versions were developed by the end of the seventeenth century.

Records describing Flamsteed's *camera obscura* seem slightly contradictory. In Francis Place's engraving of around 1676, a large square room with several windows is shown; and, in the ground-plan of the site (see p.15), the eastern summer-house is shown with a similar pattern of two windows on the south side (facing the Sun) and single windows in the two other visible walls. In 1710, German scholar Count Zacharias Conrad von Uffenbach visited the Observatory and records that Flamsteed had two cameras '… on either side of [the Octagon Room] … which are uncommonly pleasant on account of the charming prospect and the great traffic on the Thames'. Whether Flamsteed had moved his solar observatory from the summer-house or had installed two new observatories in the turrets on the roof of the house is not clear. However, the arrangement preserved in Place's engravings does not represent either of the turret rooms, which are much smaller and do not have the particular arrangement of windows evident in the print.

There is no mention of the *camera obscura* in the Observatory records for the next fifty years. It is possible that one or all were dismantled by Flamsteed's widow, when she repossessed the instruments after his death. In 1778, however, there is a notice of Nevil Maskelyne installing what is described as a 'new' *camera obscura* in the western turret of the house and paying for it himself, claiming that it added a 'fresh spur to [my] astronomical desires'. Maskelyne's *camera obscura* remained in place until 1840, when the 7th Astronomer Royal, George Biddell Airy, dismantled it and placed a new meteorological observatory in the western turret.

In 1994, the Museum installed a new *camera obscura* in the eastern summer-house. The lens has been fitted into a panning device so that visitors can experience a 120° panoramic projection of the Thames, which, even though there is less river traffic than in Flamsteed's day, remains a 'charming prospect' indeed.

48a. Hooke's 1680 *camera obscura*
Robert Hooke's work with a *camera obscura* was probably one of the inspirations behind the installation of one at the Observatory. In 1680, he presented a lecture at the Royal Society in which he showed this 'demonstration piece', a portable *camera obscura*. Pointing the small end ('A') towards an object, the members of the audience could place their faces to the hole marked 'H' and see the projected image on the viewing screen (marked 'BC').

The Camera Obscura

49a. Flamsteed's *camera obscura*, 1676
This engraving by Francis Place shows Flamsteed's solar observatory. The Latin label to the plate describes it as 'the darkened house, very convenient for receiving sunspots and solar eclipses'. The contents of the room include his 2-foot (0.6 m) refracting telescope. When the telescope was focused on the Sun, it created a 5½-inch (14 cm) diameter image. The plate upon which the image was projected was ruled with sixteen concentric circles, whose distance apart represented around 1' of arc.

49b. View of Greenwich Park by Edward Pugh, 1804
Numerous artists used Maskelyne's *camera obscura* as the basis for their picturesque views of the Thames. In around 1804, miniaturist Edward Pugh traced two views from the projection table: one facing directly down the hill to the Queen's House and the newly built West India Docks, the other towards London. Pugh's views were included in a 1805 guidebook by Richard Phillips called *Modern London*. [PU2204]

49c. View of Greenwich by Antonio Canale (Canaletto), *c.* **1753**
In the 1580s, the Italian scholar Giovanni Battista della Porta proclaimed in dismay that the *camera obscura* made it 'possible for anyone ignorant in the art of painting to draw with a pencil or pen the image of any object whatsoever'. And, over the years, there have been heated debates as to whether any of the great artists of the past used a *camera obscura* – Brunelleschi, Leonardo da Vinci, Vermeer, Canaletto. Contemporary records do describe Canaletto as using a *camera ottica*, but we have no idea what that device may have been. In any case, even if these artists did use a *camera obscura*, it was only as a starting point for the beautiful works they went on to create. [BHC1827]

49d. View of the Queen's House, 1995
This image of the Queen's House was produced in 1995, by placing a large sheet of light-sensitive paper on the imaging table of the newly installed *camera obscura*.

Daily Life at the Observatory

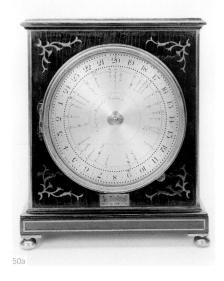

50a

How many ecstatic hours does the astronomer spend at the eye-end, high up on his scaffold-like observing chair, communing with other worlds during the darkest hours of the night? No wonder, then, that the making of colossal 'equatorials' should be replete with wondrous incident, and the details of their history almost beyond belief.
P. Fitzgerald, *Strand Magazine*, 1896

For the first 150 years of its life, the Royal Observatory staff consisted of no more than the Astronomer Royal and his two assistants. Flamsteed's most notable assistant, Abraham Sharp, was both a careful calculator (which helped to save Flamsteed himself from hours of tedious figuring) and also a skilled instrument-maker, responsible for constructing and dividing the arc of Flamsteed's 7-foot (2.1 m) astronomical sextant.

By the end of the nineteenth century, approximately forty people were working full-time at the Observatory, about half of whom were employed as calculators or 'computers'. Some assistants were great scientists in their own right. James Glaisher, for example, was a pioneer in meteorology and the main founder of the Royal Meteorological Society. He was also a pioneering balloonist who, on 5 September 1852, reached an altitude of over 30,000 feet (10,000 m), higher than any other man has attained since without the aid of oxygen.

Most of the assistants, however, led rather drab lives. Sir George Biddell Airy, the 7th Astronomer Royal, claimed that he kept the pay meagre because he did not want the majority of them hoping for a career at the Observatory. Perhaps John Pond, the 6th Astronomer Royal, summed up his own needs most succinctly:

I want indefatigable, hard-working, and, above all, obedient drudges (for so I must call them, although they are drudges of a superior order), men who will be contented to pass half their day in using their hands and eyes in the mechanical act of observing, and the remainder of the day in the dull process of calculation.

Nevertheless, the work of these 'drudges' produced the finest record of any international scientific institution.

Thomas Evans, one of the astronomical assistants under Maskelyne, described what it was like working at the Observatory in the late 1790s:

Nothing can exceed the tediousness and ennui of the life the assistant leads in this place, excluded from all society except, perhaps, that of a poor mouse … Here, forlorn, he spends his days, weeks, months in the same wearisome computations, without a friend to shorten the tedious hours, or a soul with whom he can converse … A zealous wish on his part to promote so divine a science as that of astronomy, joined to an awful contemplations of the wonderful works of the Almighty, are the sole objects that afford his pleasure in this solitary hermitage.

50a. Observatory clock by Thomas Taylor, date
Invented by Thomas Taylor, an assistant at the Observatory, this clock was designed to wake up the astronomer when particular stars were due to pass overhead. The stars are listed around the dial and flanked by small holes into which pegs were placed to activate the alarm. [D7113]

51a. An astronomer observing
An eager astronomer seated at his transit instrument with his pen and paper, ready to record the transit of a star. Note the other tools of his trade: an astronomical regulator, a barometer, a night-light and a woolly cap to keep his head warm. [B305]

Daily Life at the Observatory

51b. Reading the chronograph
In 1852, Airy installed a chronograph at the Observatory. It automatically recorded second impulses (and later two-second impulses) that had been generated by the Sidereal Standard Clock and marked these on a paper attached to a rotating drum. The time of each observation made with the Airy Transit Circle was also recorded on this paper. When the machine came into full operation in 1854, astronomers were able to have a 'printed' record of the precise time of all of their observations. The chronograph was located in a small building attached to the north-east corner of the Airy Transit Circle Room. The structure was demolished in 1993 but the chronograph and many of its original papers have survived. [A5005]

51c. Measuring the wind
This picture from the *Illustrated London News* of 31 January 1880 accompanied an article describing the work of the Observatory, focusing on the meteorological department. The scene depicts two members of staff taking readings from the anemometer on a wet and blustery evening. The instrument, set on the top of the west turret of Flamsteed House, is described in the text as 'a self-acting apparatus for continuously noting even the momentary changes in the direction and pressure of the wind'. There was also a copper pluviometer, or rain-gauge, in the turret.

51d. At home
From 1676 to 1948, Flamsteed House served as the official residence of the Astronomer Royal. The first woman to live in the building was Margaret Flamsteed, who joined her husband in Greenwich in 1692, and the first child was their niece, Ann Heming, who joined the couple in 1694. This pen-and-ink sketch is by Miss Elizabeth Smith and dates from 1838, when George Biddell Airy was Astronomer Royal. It captures the domestic nature of the back garden of the house. The shed with a sloping roof in the centre of the picture, just behind the garden wall, is one of the few visible records of Flamsteed's original meridian observatory. [B1652/23a]

51e – 51f. Measuring the sunshine
The Magnetic and Meteorological Department was established in 1840. A series of small huts was built in the Observatory grounds to house some of the instruments. Other stations were placed on the roofs of Flamsteed House. This picture shows one of the astronomers climbing up to read a paraheliometer, or sunshine recorder. The glass sphere focuses the Sun's rays on a strip of card, leaving a trail of scorch marks as the Sun moves across the sky. Exact hours of sunshine can be measured from how much of the card is burnt. Sunshine readings were taken on a daily basis at the Observatory until 2001. In the background of the picture, the time ball and the early cylindrical form of the Great Equatorial Building are visible.
[A5007 and B6899]

Into the Twenty-First Century

The office of the Astronomer Royal has continued without interruption since 1675. The Observatory itself, however, has seen perpetual change. As its work expanded, old buildings were modified and numerous, new 'temporary' structures were added, until nearly every part of the site and even some extra land in Greenwich Park was occupied by its buildings.

During the 1920s, the astronomers noticed that the newly built Greenwich power station and the electrification of the local railways were affecting their magnetic readings. By the 1930s, air and light pollution also began to hamper observations. As a result, bit by bit, the scientific work of the Observatory was moved away from London. In 1946, the Admiralty announced that the 'Royal Greenwich Observatory' (or RGO) would leave Greenwich completely for a new home at Herstmonceux Castle in Sussex. The Astronomer Royal left in 1948 and the last positional observations at Greenwich were made in 1954.

From the first hints that the astronomers might leave, the future of the Greenwich site became an issue. In 1947, the Inspector of Ancient Monuments stated that Flamsteed House, its terrace, walls and the Altazimuth Pavilion were worth saving. They also said the Victorian Physical Observatory looked like an ice-cream parlour and could be pulled down without any great loss – a fascinating insight into changing tastes. For the next ten years the Ministry of Works dithered about the site's future, the three main options being to convert it into living quarters for married members of the Royal Naval College or make it part of either the Science Museum or the National Maritime Museum. After much wrangling, the Octagon Room was put in the care of the latter 'for opening to visitors' in 1951 and, over the next few years, almost room-by-room, the Museum took possession of the site. The newly refurbished Octagon Room was opened by Prince Philip in 1953 and the whole of Flamsteed House was officially opened by Queen Elizabeth II in July 1960. The 'Greenwich Caird Planetarium' in the New Physical Observatory was opened to visitors in 1965. Finally, after further major conversion and refurbishment work, what is now called the Meridian Building was the last to be opened to the public: on 19 July 1967 by the 11th Astronomer Royal, Sir Richard Woolley. The whole complex was formally called the 'Old Royal Observatory', and was run as part of the National Maritime Museum.

In 1990, the RGO at Herstmonceux moved once more, this time to Cambridge, though most of its observing was by then done at the Northern Hemisphere Observatory on the Canary Island of La Palma. Finally, in 1998, the RGO was officially disbanded as part of a general reorganization of British astronomical work. At that point, the name of the Greenwich site was changed again, by Royal Warrant, to the 'Royal Observatory, Greenwich' (ROG).

52a. The New Physical Observatory

In 1837, the southern part of the Observatory site was enclosed. It was originally known as 'the Magnetic Ground' because it was here that the 7th Astronomer Royal, George Biddell Airy, had built up two buildings to house the Observatory's magnetic instruments and experiments.

Construction of the 'New Physical Observatory' began in 1891. Its main function was to provide much-needed office, computing and accommodation space. The architectural harmony of the current structure belies the fact that it was built in three separate phases to the design of William Crisp, an Admiralty architect, and did not take the final form of a three-storey cruciform building surmounted by a central dome until the last section was added in 1899.

The 30-foot (9.1 m) telescope dome was originally built to house the Lassell 2-foot (61cm) reflecting telescope that had been used to discover Neptune's moon in 1846 and Saturn's eighth satellite, Triton, in 1848. The Lassell reflector was later replaced by two massive new telescopes: a 26-inch (66 cm) photographic reflecting telescope and 30-inch (76.2 cm) photographic refracting telescope, both of which were presented to the Observatory in 1896 by Sir Henry Thompson, a distinguished surgeon and amateur astronomer. These two telescopes were moved to the RGO at Herstmonceux Castle in the mid 1950s and are still there. [P39986]

Into the Twenty-First Century

53c. A working observatory, c.1974
Before 1960, the old Magnetic Ground of the Observatory was cluttered with other buildings. In addition to the Physical Observatory and the Altazimuth Pavilion, there was a large Library building, two storage sheds and a stable block, as well as a number of temporary structures used to carry out specific short-term experiments. The site became so crowded that, in 1898, the Royal Parks allowed the 8th Astronomer Royal, William Christie, an additional plot of land to the south-east of the Observatory grounds in the angle of Lovers' Walk and Bower Avenue. This came to be known as 'The Christie Magnetic Enclosure' and contained seven additional buildings. All these additional structures were demolished between 1956 and 1958 when the astronomers moved to Herstmonceux. [B1006-5]

53d. Frank Carr, 1903–91
Negotiations to secure the Observatory site and collections for posterity spanned the careers of two directors of the National Maritime Museum. Geoffrey Callender, a naval historian, was a key figure in creating the Museum and was its first director from 1934. It was he who first began unofficial talks with the 10th Astronomer Royal, Harold Spencer Jones, in the early 1940s. When Callender died in 1946 (still in office aged 70), his successor Frank Carr took up the baton with a renewed and personal interest, since both his grandfather and brother had been astronomers at the Observatory. Over the next decade, however, the escalating costs of securing and refurbishing the Observatory produced considerable frictions between the Ministry of Works and the Museum's Board of Trustees. Partly as a result, Carr was forced into early retirement in December 1966, just seven months before the full public opening of the Observatory in July 1967 – a ceremony to which, sadly, he was not invited. [4592-1]

53a. Astronomia
The building's striking exterior terracotta decoration was designed by W. J. Neatby and cast by Doulton & Co. at its local Lambeth factory in 1895. Over the main, north door is a bust of John Flamsteed, the 1st Astronomer Royal, and decorative plaques bearing the names of famous astronomers, scientists and instrument-makers are woven into the cornices of the building. At ground level, on the north-west side, is a terracotta relief depicting *Astronomia*. With her arms outstretched across the heavens, she holds the Sun and the Moon in her hands. A portion of the ecliptic, or zodiacal band, can be seen curving behind her back, which bears images of the constellations of Aquarius, Scorpio, Leo and Taurus. [F6911-060]

53b. The Altazimuth Pavilion
Whereas Airy had maintained that 'the Observatory was not the place for new physical investigations', the 8th Astronomer Royal, William Christie, was fully committed to the new science of astrophysics, which focused specifically on the physical make-up of the heavenly bodies. The tiny Altazimuth Pavilion was completed in 1899 as part of Christie's campaign to expand the Observatory's research in this field. The name 'Altazimuth' comes from the mounting of the telescope that was originally housed in the building. An altazimuth mounting has two independent rotational axes which allow the telescope to move from the horizon to the zenith (altitude) and around the circle of the horizon (azimuth). It is a very useful mounting for tracking objects as they move across the sky. The Observatory suffered considerably in World War II and the most visible remaining scar is the south cornice of this building. It was originally in terracotta, matching the north one, but was blown off by a bomb in 1944 and repaired more simply in brick. [F6935-034]

The Observatory Today

Today, the Observatory at Greenwich combines all that is best of old and new. The historic buildings have been expertly restored and sensitively adapted where necessary, and several of the Greenwich telescopes and clocks have returned to their original locations in the Meridian and Great Equatorial Buildings, giving visitors an insight into what these spaces looked like when the astronomers were still in residence. The buildings at the north end of the site also hold exciting displays on the history of astronomy and timekeeping. Visitors can stand astride the Prime Meridian of the World, with one foot in the eastern hemisphere and the other in the western, and see the Airy Transit Circle, the telescope used to define Greenwich Mean Time. In Flamsteed House, they can also view the four great marine timekeepers built by John Harrison in the eighteenth century, which are still in perfect working order.

The fine late-Victorian buildings at the south end of the complex have now also been fully renovated, with special attention to preserving and restoring their delicate terracotta mouldings. Inside, however, both have been remodelled to create the new Weller Astronomy Galleries, which explore the story of time and space in the twenty-first century. And between these and the Altazimuth Pavilion, the new 120-seat Peter Harrison Planetarium has been constructed so that everyone who visits can have a clear view of the beauties and wonders of the night sky.

54a

54a. Purpose-built for Greenwich
The exterior shell of the the Peter Harrison Planetarium reflects a collaborative design formulated by former RGO astronomer Dr Robin Catchpole and the architectural firm of Allies and Morrison. It reflects the history of the Royal Observatory as well as celebrating its unique location on the surface of the Earth. The truncated-cone structure embodies four precise sets of astronomical data:

- The main axis of the cone is set north-south, along the local meridian of the site and slightly to the east of the Prime Meridian of the World.
- The inclined angle of the south side of the cone is specific to the latitude of the Royal Observatory of 51° 28' 39 – so that its projected tip points directly towards the northern celestial pole, or the Pole Star, which is located in the tail of the constellation of Ursa Minor (the Little Bear). The sighting line in the cone's south side points directly towards this star.
- The northern side of the cone is aligned with the local zenith, or the point in the sky that is perpendicular to the local horizon of Greenwich. This angle is also unique to the latitude of Greenwich.
- The cone has been truncated at an angle lying parallel to the plane of the Equator. The angle between this plane and the inclination of the cone (the celestial axis of the poles) is exactly 90°. This intersection of local and celestial planes is also specific to Greenwich's location.

The bronze surface of the cone itself was hand-patinated by Andy Elton. Its deep blues, greens and browns not only harmonize with the local setting of Greenwich Park, but are also intended to evoke the nebulous stellar dust of deep space. [F6514-073]

55a. Seeing Spots
From 1873 onwards, astronomers at Greenwich used a photoheliograph to study the surface of the Sun. The Sun's rays are so bright that any direct viewing can cause permanent damage to the retina of the eye by burning it. Astronomers solved this problem by finding ways to project the Sun's light onto a blank surface using devices such as the *camera obscura*. With the invention of photography, astronomers developed the photoheliographic telescope: a specially adapted telescope in which the gathered light is magnified and focused on a photographic plate. Today, there are special filters that can be used to block the damaging rays of the Sun.

The Sun is our local star, which warms us and provides energy for all forms of life on Earth; without it we would not and could not exist. By studying it, astronomers are also able to understand more about the nature of stars in general: their composition and histories, as well as their possible futures. The Sun has no permanent features because it is gaseous. It is mostly incandescent hydrogen, which is so compressed at the Sun's core that it fuses into helium – exactly the same thing that happens when a hydrogen bomb is exploded. This constant reaction results in the Sun converting 240 million tonnes of mass into energy every minute. Since the Sun is constantly in a volatile state, its different layers are always subject to disturbances. Prominences and solar flares – huge eruptions from the corona that extend for tens of thousands of kilometers – as well sunspots, are all indicators of disturbances in the Sun's magnetic field. Their effect can be experienced on Earth as they regularly cause disruption to radio communications, knock satellites off-course, and even cause extreme changes in normal weather patterns. By analysing sunspot activity on the Sun's surface, astronomers have been able to detect an eleven-year cycle in the Sun's volatility, and thereby help to understand more about how these changes in the Sun affect life on Earth. [E0443-2]

The Observatory Today

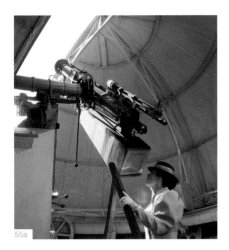

55b. Weller Astronomy Galleries
In 2007, the Museum re-opened the southern part of the Observatory to visitors with a number of new facilities designed to bring the Observatory story up to date. Here, the visitor can learn about recent developments in modern astronomy and explore mankind's ongoing quest to understand the mysteries of the universe. At the heart of the new complex is the Peter Harrison Planetarium, where a practising astronomer leads visitors on a personal journey through time and space, using state-of-the-art technology. The new Weller Astronomy Galleries tell the story of how modern astronomers study the universe, what they look for and how this work informs what we think we know about the universe today. And, in the tiny Altazimuth Pavilion, a working array of historic and modern telescopes offer pre-booked visitors the opportunity to study the surface and atmosphere of the Sun. [F6935-025]

55c. The largest telescope in Britain
The 28-inch (71 cm) refracting telescope, housed in the onion-shaped dome in the Great Equatorial Building, is the largest of its kind in Britain and still the seventh-largest telescope in the world. It was commissioned in 1885 by the eighth Astronomer Royal, William Christie. The 28-inch diameter lens was designed by renowned Dublin-based optical manufacturer Howard Grubb, and the job of casting was awarded to the Chance Brothers of Birmingham in 1888. It took fifteen attempts and three-and-a-half years before two blanks were successfully cast and ready for polishing. The completed lens, weighing 200 lbs (90.7 kg), was finally installed in 1893. Until 1947, the telescope was used almost exclusively for observing double-stars, a pair of stars that orbit around each other because they are held together by their mutual gravitational attraction. It was dismounted in 1947 and transferred to Herstmonceux, where it continued its working life until the mid-1960s. The telescope was returned to Greenwich and re-erected on its original mount in time for the tercentenary of the Observatory in 1975.
The present 'onion dome' also dates from that time, and is a fibre-glass replica of the 1894 iron-framed dome, which was severely damaged in World War II and later removed. Today, the 28-inch refractor is still in perfect working order. There are regular viewing sessions for visitors to the Observatory, mostly on late-autumn and winter evenings, when the skies darken earlier. For additional information about viewing the night sky through the 28-inch refractor, check the web pages on www.nmm.ac.uk. [E9663-3]

55d. Star-gazing
Even though the light pollution of London makes the conditions for star-gazing at Greenwich less than ideal, it is certainly possible to view the features of the Moon, the rings of Saturn or the moons of Jupiter through one of the Observatory's telescopes. There are also regular, safe-viewing sessions of sunspots, using the Observatory's portable telescopes. During solar and lunar eclipses, the Observatory often runs special events, with trained astronomers on hand to explain what is happening as it occurs. The Observatory also publishes numerous helpful fact-sheets on its website for those who cannot come to Greenwich but want to learn more about astronomy. It offers advice about buying a first telescope and provides information on what is currently visible in the night sky. There is also a link to information about the first appearance of the new crescent moon for those waiting to celebrate the festival of *Eid ul-Fitr* following the Islamic holy month of Ramadan. For more information on the astronomical information provided by the Observatory, check the web pages on www.nmm.ac.uk. [E0459-17]

Everything You Wanted to Know
About the Universe

56a

The content of the displays in the Weller Astronomy Galleries has been assembled by the Observatory staff in close collaboration with a number of practising professional astronomers who are doing active research within a wide variety of specialized disciplines. Working together, they have tried to provide insights into some of the topics that interest them but which also have a deep resonance in everyone's mind:

• What is our place in the universe?
• What are the fundamental keys that astronomers use to unlock the secrets of the universe?
• What do astronomers think they know about the universe right now?
• What is there left to discover?

56a. The Gibeon meteorite
A meteor is a piece of space rock, usually a small piece of a comet or asteroid, that enters the Earth's atmosphere. As it falls, air friction causes it to ignite, producing the spectacular fireworks of what is known as a 'shooting star'. Few meteors survive their passage through the atmosphere; those that do are called 'meteorites'. It was not until 1803 that scientists accepted that meteorites did, indeed, fall from space.

57a. The big ideas
Modern astronomers study the stars for exactly the same reasons that ancient ones did: they want to understand the history of the universe and all that entails. How and when was the universe formed? Where does life come from? How big is the universe – is it finite or infinite, expanding or contracting? What does the future hold for humans, for our solar system, for our Sun?

Astronomy is like any other science – based on theorems, experimentation and proofs – except that deep space cannot be brought into the laboratory for close examination. Instead, astronomers have to rely on what they can glean from the light that is being transmitted from the stars to Earth, and of their understanding of the different factors that might effect a change in the quality of the light itself. Astronomers can analyse starlight to learn about the chemical composition of the stars. They can determine if a star is moving away from Earth or towards it. They can tell how old a star is by the light it emits. They can even tell if the light has been bent by some unseen gravitational force, such as a black hole.

In the vacuum of deep space, light travels at just over 186,000 miles (or nearly 300,000 km) per second. Since the Sun is about 93 million miles (150 million km) from Earth, the light from the Sun takes a little over eight minutes to reach us. The light from the second-closest star to Earth, *proxima Centauri*, takes about 4.2 years to reach us. And, by using extremely sensitive telescopes, astronomers can now detect starlight that has taken over 13 billion years to reach the Earth from distant galaxies. Put another way, when this light reaches the Earth, it is more than 13 billion years old and may come from stars that no longer exist. To look at the stars, then, is to look back into the history of the universe.

57b. Brass laboratory spectroscope made by John Browning, London, 19th C.
The main function of a telescope is to collect light. The simplest telescopes collect visible light, but visible light forms only a small part of a larger spectrum of radiation. Radio telescopes, for example, collect invisible light waves emitted by very cold gas clouds in deep space. Infrared telescopes can 'see through' dark clouds of dust and gas and reveal the starlight hidden behind them. X-ray space telescopes detect extremely hot gases, while microwave telescopes can pick up the faintest bits of light left by the explosion of the 'Big Bang'. [D1214-1 and F6935-014]

57c. Newsflash!
Astronomy is a living science, with discoveries being made all the time. To keep visitors up to date with the latest breaking news, the displays in the Weller Astronomy Galleries contain a news kiosk that displays information about current topics and recent discoveries. Additional discussions of astronomical stories in the news can be found by logging on to the Museum's website: www.nmm.ac.uk.
[F6947-022]

57d. Ask an astronomer
In an attempt to ensure that information conveyed at the Weller Astronomy Galleries is topical and factually correct, the content of the displays has been devised by practising professional astronomers – many of whom are working at the cutting edge of their profession. The team includes astronomers working on the solar system, extra-solar planets, planetesimals (bodies of rock and/or ice less than 6.2 miles or 10 km in diameter), planetary probes, solar physics, star formation, galaxy formation, black holes, cosmological simulations and gravitational lensing (when the path of light rays from a distant light source have been bent by the gravitational pull of a massive object between it and the Earth). Some of the astronomers agreed to be filmed for the displays – discussing different aspects of their work and explaining the significance of their current research. [F6935-007]

57e. Seeing the light … and more
For over a century, astronomers have been able to study the chemical composition and relative temperature of the stars by means of spectroscopy. As every schoolchild learns, so-called 'white light' is actually made up of a fused spectrum of coloured lights: red, orange, yellow, green, blue, indigo and violet. In the early nineteenth century, the German optician Josef Fraunhofer examined the spectrum created by the light coming from the Sun and was surprised to discover that there were a number of dark lines crossing it. In 1859, another German scientist, Gustav Kirchhoff, discovered the significance of Fraunhofer's lines. He realized that they were produced by the gaseous chemicals in the cooler upper layers of the Sun, which absorbed the light

Everything You Wanted to Know About the Unvierse

from this particular segment of the spectrum. His later researches led to the discovery that each chemical produces its own unique pattern of lines. A spectroscope uses specially designed prisms to break down the 'white light' coming from a star into an extremely detailed spectrum. By analysing the 'Fraunhofer lines', an astronomer can determine the chemical components of the star. The star's temperature can also be measured by the intensity of the individual lines. Originally an astronomer would look through a spectroscope attached to his telescope and record his findings in a drawing. Today, this information is recorded with an electronic light sensor called a 'charge-coupled device, or 'CCD'. It uses the same technology as digital cameras but is 200 times more sensitive. [D6935-1]

57f. Into outer space

In 1957, the former Soviet Union launched *Sputnik 1*, the first artificial satellite into space. Since that time, numerous unmanned space probes have been sent into outer space, gathering data and transmitting this raw information back to Earth, where it is sifted and analysed by astronomers eager to test their theories. The *Beagle 2* lander was sent to find signs of life on Mars and was expected to land on Christmas Day 2003. Unfortunately, after the probe left the *Mars Express* spacecraft, controllers were unable to re-establish contact with the lander, so its landing has not yet been confirmed. Space exploration in the 1960s and 70s was marked by intense competition. The so-called 'space race' was conducted mainly between the United States and the Soviet Union, each of which saw political and military advantages in winning. Today, however, most space projects are the result of international collaboration, combining the best scientific knowledge and technological expertise of several countries to produce the best product.

The Peter Harrison
Planetarium

When the original forty-seat Caird Planetarium opened in the Lassell Dome on the south side of the site in 1965, the third director of the Museum, Basil Greenhill, reported that the success of the project was 'almost embarrassing': as the staff were overwhelmed by the positive public response. Over the years, as the Observatory's visitor numbers grew, so did the popularity of the old planetarium – and the embarrassment of not being able to satisfy all those who wanted to enjoy it.

In spring 2004, the Museum launched a campaign to help fund a new planetarium for the Observatory. The plan was to build a 120-seat facility with a state-of-the-art projection system. As has always been the case in Greenwich, this would be supported by live commentary from experienced members of staff.

In November 2005, the National Maritime Museum was pleased to announce that Peter Harrison had agreed to fund the planetarium development, through the Peter Harrison Foundation 'Opportunities for Education' programme. Mr Harrison said that the new planetarium would 'provide a remarkable educational tool for generations of future visitors'.

58a. Bringing the stars to you

Over 99% of the people living in Europe and the United States are affected by the light pollution caused by street lamps, cars, buildings, and so on. The indiscriminate spilling of all this light into the night sky masks the starlight to such an extent that most people can no longer see the outline of the Milky Way with the naked eye. In London, for example, it is impossible to see any but the very brightest stars. Many people grow up never having experienced the true magic of the night sky or being able to behold the grandeur of the universe as it slowly wheels above their heads.

Planetariums were originally devised as educational tools – to help people understand the physics behind the apparent daily and yearly movements of the heavens. Now, however, they are one of the few means by which most people will ever be able to appreciate the nightly spectacle that our ancestors took for granted.

Many people, both old and young, come as part of a pre-booked educational programme. Astronomy is increasingly seen as a 'user-friendly' science, providing first-level access to a number of other scientific disciplines, such as mathematics, physics, chemistry and biology. The Museum runs a series of formal programmes for school groups from primary through to university level. It also organizes more informal workshops on astronomy as an inspiration for creative writing in both prose and poetry, as well as hosting sessions for those interested in the visual arts. Information on these programmes can be found on the museum's website www.nmm.ac.uk.
[6935-001]

59a – 59b. Recreating the heavens

A planetarium is based on the principle of recreating the effect of the night sky by projecting the exact pattern of the stars on the inner surface of a curved dome. The projector used in the the Peter Harrison Planetarium is a Digistar 3 Laser system made by the American firm, Evans & Sutherland. It is capable of displaying over 16 million pixels of video, which enables it to simulate the colour and clarity of the real stars in the night sky. The projector's light source uses red, green and blue (RGB) laser diodes and fibre-optic amplifiers from the telecommunications industry, and it operates on standard 13-amp outlets, consuming less that 1kW – so it is economical and environmentally friendly as well. The curved dome was specially manufactured for the planetarium by Spitz, a company which has been making observatory domes in Pennsylvania since 1946. Since the first planetarium lectures at Greenwich in 1965, the Observatory has remained committed to having a live presenter deliver each planetarium show. Not only is the content of every programme devised by the Observatory's own in-house team of astronomers and education specialists, but the added benefit of having a knowledgeable presenter is that every show can be adapted to meet the special interests and abilities of each visiting group.
[F6844-010 and F6947-012]

The Peter Harrison Planetarium

Astronomers Royal:
Past and Present

John Flamsteed
(1646–1719)
1675–1719

Edmond Halley
(1656–1742)
1720–42

James Bradley
(1693–1762)
1742–62

Nathaniel Bliss
(1700–64)
1762–64

Nevil Maskelyne
(1732–1811)
1765–1811

John Pond
(1767–1836)
1811–35

Sir George Biddell Airy
(1801–92)
1835–81

Sir William Christie
(1845–1922)
1881–1910

Sir Frank Dyson
(1868–1939)
1910–33

Sir Harold Spencer-Jones
(1890–1960)
1933–55

Sir Richard Woolley
(1906–86)
1956–71

Sir Martin Ryle
(1918–84)
1972–82

Sir Francis Graham-Smith
(b. 1923)
1982–90

Sir Arnold Wolfendale
(b. 1927)
1991–94

Lord Rees of Ludlow
(b. 1942)
1995–current

Suggested Further Reading

W. J. H. Andrewes (ed.)
The Quest for Longitude: The Proceedings of the Longitude Symposium
Harvard University, Cambridge, Massachusetts, November 4-6 1993 (Cambridge, MA, 1993)

Clive Aslet
The Story of Greenwich (London, 1999)

Jonathan Betts
Harrison (London, 2007)

Time Restored (Oxford, 2006)

Pip Brennan
The *Camera Obscura* and Greenwich (London, 1994)

A. Chapman (ed.)
The Preface to John Flamsteed's *Historia Coelestis Britannica* or British Catalogue of the Heaven (1725), transl. by A.D. Johnson (London, 1983)

Gloria Clifton
Directory of British Scientific Instrument Makers, 1550-1851 (London, 1995)

Gloria Clifton and Nigel Rigby, (eds.)
Treasures of the National Maritime Museum (London, 2005)

Elly Dekker et al.
Globes at Greenwich. A Catalogue of the Globes and Armillary Spheres in the National Maritime Museum (Oxford, 1999)

Eric G. Forbes, Derek Howse and A.J. Meadows
Greenwich Observatory, 3 volumes (London, 1975)

R. T. Gould
'John Harrison & his Timekeepers', The Mariner's Mirror, XXI, no. 2, 1935
(Reprinted as a separate fascicule by the NMM, London, 1978, with several reprints.)

Hester Higton et al.
Sundials at Greenwich. A Catalogue of the Sundials, Horary Quadrants, and Nocturnals in the National Maritime Museum (Oxford and London, 2002)

Derek Howse
Francis Place and the Early History of the Greenwich Observatory (New York, 1976)

Greenwich Time and the Discovery of Longitude
(Oxford, 1980; 2nd edn: *Greenwich Time and the Longitude*, London, 1997)

Nevil Maskelyne, the Seaman's Astronomer (Cambridge, 1989)

Anthony Jones
**Splitting the Second:
The Story of Atomic Timekeeping** (London, 2000)

Henry C. King
The History of the Telescope
(High Wycombe, Bucks, 1955)

Kristen Lippincott
Astronomy (London, 1994)

A Guide to the Old Royal Observatory, London 1995.
Reprinted as *A Guide to the Royal Greenwich Observatory: The Story of Time and Space* (London, 2002 and 2005)

Kristen Lippincott et al.
The Story of Time
(London, the National Maritime Museum, 1 December 1999 – 28 September 2000; London, 1999)

Kevin Littlewood and Beverley Butler
Of Ships and Stars: Maritime Heritage and the founding of the National Maritime Museum, Greenwich (London and New Brunswick, NJ, 1998)

Stuart Malin and Carole Stott
The Greenwich Meridian (Southampton, 1984)

Walter Maunder
The Royal Observatory Greenwich: A Glance at Its History and Work (London, 1900)

William Hunter McCrea
Royal Greenwich Observatory: An Historical Review Issued on the Occasion of Its Tercentenary (London, 1975)

**The Old Royal Observatory:
The Story of Astronomy and Time**
(London, [n.d., 1990])

**The Old Royal Observatory, Greenwich:
Guide to the Collections** (London, 1998)

The Oxford Dictionary of National Biography (Oxford, 2004)

Colin A. Ronan
Edmond Halley: Genius in Eclipse
(London, 1970)

Colin A. Ronan (ed.)
Greenwich Observatory: 300 Years of Astronomy (London, 1975)

Dava Sobel
Longitude: The True Story of a Lone Genius Who Solved the Greatest Scientific Problem of His Time (New York, 1995)

Gerard L'E. Turner
Antique Scientific Instruments (Poole, Dorset, 1980)

Gerard L'E. Turner
Elizabethan Instrument Makers: The Origins of the London Trade in Precision Instrument Making (Oxford, 2000)

Koenraad van Cleempoel et al.
Astrolabes at Greenwich: A Catalogue of the Astrolabes in the National Maritime Museum (Oxford and London, 2006)

Christopher Walker (ed.)
Astronomy Before the Telescope (London, 1996)

Peter Whitfeld
The Charting of the Oceans: Ten Centuries of Maritime Maps (London, 1996)

Emily Winterburn
The Astronomers Royal (London, 2003)

Credits

For all enquiries contact:
National Maritime Museum, Greenwich,
London SE10 9NF
Telephone: 020 8858 4422
Fax: 020 8312 6632

Or visit our websites:
National Maritime Museum:
www.nmm.ac.uk
Royal Observatory Greenwich:
www.rog.nmm.ac.uk

Opening times:
10.00 – 17.00 hours daily; last admission 16.30 hours. (Please enquire for dates of summer opening.) Closed 24-26 December inclusive.

Photography is not permitted inside the buildings. Please check in advance for changes to opening dates and times by calling information on +44 (0)20 8312 6565. To book a group visit please call 020 8312 6608 or e-mail: bookings@nmm.ac.uk

Photographic acknowledgments
Images from the NMM collection may be ordered by writing to the Picture Library, National Maritime Museum, Greenwich SE10 9NF or online at www.nmm.ac.uk. Please quote the image number provided alongside the relevant image text. All images copyright © National Maritime Museum, London

Illustrations are also reproduced by kind permission of the following:

Ägyptisches Museum und Papyrussammlung, Staatliche Museen zu Berlin/© BPK, Berlin – Margarete Büsing: **8b**
All Rights Reserved Beagle 2: **57h**
BT Archives: **35e**
Courtesy of David Penney: **29c**
Courtesy of Jonathan Betts: **27f**
Department of Special Collections, Skidmore College, Saratoga Springs, New York: **45c**
Derby Museum & Art Gallery/© AKG Images, London: **21d**
English Heritage Photo Library: **9a**
Illustrated London News: **51c**
Kyongju City, Kyongsang Province, South Korea/Bridgeman Art Library **9b**
MSFC/Chandra X-ray Center, Harvard/NASA/CXC/SAO: **58a**
NASA Headquarters – GReatest Images of NASA (NASA-HQ-GRIN): **57a**
National Portrait Gallery, London: **12a, 13a**
Royal Greenwich Observatory, Cambridge: **15b, 15c, 40a, 43d, 47d, 48a, 49a**
The Royal Society, London: **14a**
Science Museum/Science & Society Picture Library: **21c, 28, 31c, 32a, 35d, 41c, 47b**
Tobias Mayer Museum, Ludwigsburg: **31b**
The Worshipful Company of Clockmakers' Collection, UK/ Bridgeman Art Library, London **33c, 33d**

© 2007 National Maritime Museum.
All rights reserved. No part of this publication may be reproduced, stored in a retrieval system, or transmitted, in any form or by any means, without prior permission of the publishers and copyright holders.

ISBN 13: 978 0 948065 82 8

Text by Kristen Lippincott
Design by Marazzi Jones
Photography by Josh Akin, Ben Gilbert, Ken Hickey, Andrew Holt and Tina Warner
Picture credits – National Maritime Museum unless otherwise stated
Picture research by Sara Ayad
Project Management by Abigail Ratcliffe
Production Management by Geoff Barlow
Printed in Italy on FSC certified paper by Printer Trento Srl

Certificate No. CQ–COC–000012

The paper used for this book has been independently certified as coming from well-managed forests and other controlled sources according to the rules of the Forest Stewardship Council.

This book has been printed and bound in Italy by Printer Trento S.r.l., an FSC accredited company for printing books on FSC mixed paper in compliance with the chain of custody and on-products labelling standards.